A Foundation In Precalculus

For my parents. I will never be able to repay your lessons on hard work, integrity, and perseverance.

Typset and published by Asymptotic Group LLC.

2020

The publisher makes no claims as to the accuracy of the material presented in this text. In reading this text, you agree not to hold the publisher or author liable for any damages that may result. This text may not be duplicated or reproduced in any manner without the written consent of the author and publisher.

Contents

Chapter 1. Odds and Ends	1
1.1. Beginning Thoughts	1
1.2. Prerequisites	2
1.3. A Very Brief Introduction to Logic	3
1.4. A Bit About Sets	4
Chapter 2. Function Theory	8
2.1. The Definition of a Function	8
2.2. Equality, 1-1, and Onto	14
2.3. Function Composition	18
2.4. Invertible Functions	22
Chapter 3. An Introduction to Real Functions	25
3.1. A Starting Point	25
3.2. Fundamental Functions	30
3.3. The Domain of a Function	40
3.4. Even, Odd and Equality	47
3.5. Vertical Transformations	52
3.6. Horizontal Transformations	58
3.7. Vertical and Horizontal Transformations	66
3.8. Composition	69
3.9. Composition and Graphing	75
3.10. Inverses	79
Chapter 4. An Introduction to Mathematical Modeling	87
4.1. The Least Squares Regression Line	87
4.2. Maximizing and Minimizing	96
4.3. Exponential Growth and Decay	106
Chapter 5. Exponential and Logarithmic Functions	116
5.1. Exponential Functions	116
5.2. Logarithmic Functions	120
5.3. Graphing Logarithmic Functions	127
5.4. Solving Equations With Logarithms	132
Chapter 6. Polynomial, Rational, and Power Functions	138
6.1. Polynomial Functions	138

6.2.	Rational Functions	146
6.3.	Power Functions	151

Chapter 7. An Introduction to Trigonometric Functions — 155
7.1.	Radians, Sine and Cosine	157
7.2.	The Unit Circle	165
7.3.	Combinations of sine and cosine	172
7.4.	Graphing Sine and Cosine	176
7.5.	Graphing Tangent and the Rest	184
7.6.	Vertical Transformations of Trigonometric Functions	188
7.7.	Horizontal Transformations of Trigonometric Functions	192
7.8.	Inverse Trigonometric Functions	196
7.9.	Compositions Involving Inverse Trigonometric Functions	205
7.10.	Extending to Non-Unit Circles	213

Chapter 8. Trigonometric Equations and Oblique Triangles — 218
8.1.	Trigonometric Equalities and Inequalities	218
8.2.	Trigonometric Identities	224
8.3.	The Law of Cosines	230
8.4.	The Law of Sines	236

Chapter 9. Answers To Exercises — 243

CHAPTER 1

Odds and Ends

1.1. Beginning Thoughts

The seemingly rigid nature of mathematics often results in a gross oversimplification. Namely, that correct answers are the goal. They are not. The goal is far more important and far more general. Mathematics is a process, a series of steps, reasoning, and conclusions. If your steps, reasoning, and conclusions are sound, correct answers will be an almost trivial byproduct. The goal in mathematics is understanding of this process. To understand the "why" behind the logical manipulation.

You need not worry if such an approach to mathematics is new to you. It is extremely common, and extremely unfortunate, that fundamental mathematics is often taught without an appropriate amount of "why", and is instead unnaturally forced with notions of "do this when you see that", or "remember this because it will be on a test." Adjusting your thoughts on mathematics towards understanding the bigger picture will result in a much more enjoyable experience.

Mathematics textbooks are often utilized only as a repository for formula and homework problems. You must break this trend. Do not scan through the pages searching for boxed formula, or important bold phrases. In this text, you will find neither. Instead, read everything on all of the pages. You must start out slowly. Reading at your normal pace will almost certainly result in being overwhelmed. Read one line, ask yourself if you understand its content. If you do, go on to the next. If you don't, re-read the line and several of the ones that came before it. Repeat.

This particular text has been designed on two assumptions. First and foremost, that it will be read, in entirety, from beginning to end. Consequently, each successive section will assume an understanding of the preceding ones. Be sure that you are comfortable with a topic before proceeding to the next.

The second assumption is that all of the exercises will be completed and understood. Unlike the vast majority of math textbooks, there are not hundreds of exercises at the end of each section. Rather, there are enough exercises for the reader to practice, develop, and explore the ideas that have been presented. A full understanding of several exercises is preferable against little to no understanding of many exercises. Work to understand, and to explain the process of solving all of the exercises at the end of each section.

1.2. Prerequisites

The study of precalculus requires a good working knowledge of the material that comes before it. In the case of this textbook, basic geometry, and the material covered in a traditional algebra I and II sequence are prerequisites. Lack of proper preparation is the most common reason for difficulty in precalculus.

If you find yourself a bit unprepared, all is not lost. Everyone who works with mathematics finds themselves in that position at one point or another. Realizing you need a bit of brush up work is half of the battle. The other half is actually brushing up. Thus the purpose of this section, and the review problems contained herein.

This brief review is not intended to serve as an instruction for the material presented; it is a review. If you can correctly complete these exercises, you are likely in a good place to start this text. If you find difficulty in any of these review exercises, you should seek out an instructional source (textbook, teacher, tutor, etc.) that can explain the material in depth. Now is the time to straighten out any areas of concern in the review material. Waiting to do so will only lead to the frustrating combination of trying to sort out old material while learning new material.

1.2.1. General Algebra Review.

(1) Simplify each expression. Use only positive exponents in answers.

(a) $5^{-1} + 6^{-1}$

(b) $\dfrac{(3^{-1}x^{-3}y)^{-1}(2x^2y^{-3})^2}{(5x^{-2}y^2)^{-2}}$

(c) $-2^0 + (-2)^0 - 1^0 - (-1)^0$

(d) $\left(\dfrac{zn^{-2}p}{z^2np^4}\right)^{-2} \left(\dfrac{zn^{-2}p}{z^2np^4}\right)^{3}$

(2) Solve for x. $\quad \dfrac{3}{x^2+x-2} - \dfrac{1}{x^2-1} = \dfrac{7}{2x^2+6x+4}$

(3) Solve for q. $\quad \dfrac{3}{k} = \dfrac{1}{p} + \dfrac{1}{q}$

(4) Solve for x. $\quad \sqrt{8x+8} - 1 = 2x + 1$

(5) Solve this equation two ways. First use the quadratic formula, then complete the square. $2x^2 - 7x - 9 = 0$

(6) Solve this equation two ways. First use the quadratic formula, then complete the square. $-x^2 + 4x = 1$

(7) Given the equation
$$y = -x^2 - 4x - 3$$
find the vertex, x-intercepts, and y-intercept. Then sketch a graph.

(8) Use substitution to solve the following linear system.
$$3y - 5 = 2x$$
$$4x = 2 - 3y$$

(9) Find the distance between the points $(3,5)$ and $(-2,7)$.

1.3. A Very Brief Introduction to Logic

Throughout this text (and ostensibly every other math book you will come across) words and statements that you are familiar with may not hold the meaning that you expect. Mathematicians use the logical meanings of words and phrases, and these meanings are not necessarily the same as common vernacular. As our first example, let us consider the word *or*.

Suppose you find yourself having a conversation with a friend and you say "I'll go to the store or to a movie." The common understanding of what you told your friend is that you will choose one of the following two options:

(1) Go to the store
(2) Go to a movie

It is not commonly assumed that you would do both of these choices (go to the store, and then go to a movie), because in common English, "or" is generally used in an exclusive sense. You would exclusively choose between the two options, in this case going to the store versus going to a movie.

In mathematics, "or" is not used exclusively. Instead, it is used inclusively. The same statement, "I'll go to the store or to a movie" now allows one of three possible options:

(1) Go to the store
(2) Go to a movie
(3) Do both

Performing any of these three options would make your statement true.

Another word that is frequent in both mathematics and common English is *and*. Fortunately, "and" carries the same meaning in both cases. That having been said, it is still worth our time to consider an example.

Suppose you tell your friend, "I'll bring some cheese and crackers to the party." If you show up to the party with just cheese, then your statement is false (and your friend is mad at you for ruining a perfectly good party). If you show up with just crackers, then again your statement is false (and your friend begins to wonder why you lie all the time). In order for your statement to be true, you must two items to the party:

(1) Cheese

(2) Crackers

Another bit of phrasing that can bring about trouble is "if then", also known as a *conditional* statement. Consider the statement, "if you add two even numbers, then the sum of those two numbers is even." This statement is true in exactly one direction, namely the one that it has been presented in. If two even numbers are added, then the sum is even.

It is an extremely common misstep to try and use conditional statements in the reverse direction. Let us consider our earlier statement in reverse. Suppose you knew that the sum of two numbers was even. Could you conclude that the two numbers that had been added were themselves even? You cannot. While two even numbers add to an even number, two odd numbers also add to an even number. Knowing that the sum is even does not guarantee that the two numbers were themselves even.

Similar to a conditional statement, a *biconditional* statement uses the phrasing "if and only if." However, unlike a conditional statement, a biconditional statement works in both directions. Consider the statement, "the product of two numbers is odd if and only if both of those numbers are odd." This statement is true in both directions. If you know that two numbers are odd, then their product must be odd. If you know that the product of two numbers is odd, then both of those numbers must be odd.

1.4. A Bit About Sets

In mathematics, a *set* is a collection of items. There are no predetermined rules as to what these items must be. Sets can consist of numbers, variables, words, pictures, names, places, or anything else that you might desire. The collection of

$$\text{Tree}, \ 3, \ -472, \ \$, \ \clubsuit$$

is a set. It's a somewhat odd and random collection of items, but it is a set nonetheless.

Naming sets is commonplace, and is generally done with capital letters. Let us name our set from above S. When writing sets, we enclose the items inside a pair of curly brackets. Our set from above can be written

$$S = \{\text{Tree, 3, } -472, \$, \clubsuit\}$$

To indicate that a particular item is in a set - say we wanted to express that $\clubsuit$ is in our set S - we would say that, "$\clubsuit$ is an element of S." Writing out the phrase "is an element of" over and over again would get a bit tedious, so we use the shorthand notation

$$\clubsuit \in S$$

where the $\in$ is read "is an element of".

To express that a particular item is not in a set, we would say, "is not an element of." The item $\star$ is not in our set S, so we would say "$\star$ is not an element of S". The shorthand notation is written

$$\star \notin S$$

We defined our set S by explicitly writing out all of the elements in it. While this approach works well when a set contains relatively few elements (our set has only five elements), it would be impractical for sets that contain large numbers of elements. Suppose we want to define a set called T that consists of all the whole numbers from 1 to 100. Writing out all of these numbers would be a lengthy process. To avoid this, we can define T with *set builder notation*. Set builder notation is an efficient way of describing all of the elements inside a set at once.

$$T = \{t \mid t \text{ is a whole number and } 1 \leq t \leq 100\}$$

This set builder notation is read "T is the set of all things t such that t is a whole number and t is greater than or equal to 1 and less than or equal to 100." The lower case t represents the way we refer to a generic, unspecified element of the set T. The vertical bar, $|$, is read as "such that."

As another example, let us create a set called D that consists of the days of the week. Writing out D explicitly would give us

$$D = \{\text{Monday, Tuesday, Wednesday, Thursday, Friday, Saturday, Sunday}\}$$

Using set builder notation we would have

$$\underbrace{D = \{d \mid}_{D \text{ is the set that of all things } d \text{ such that}} d \text{ is a day of the week}\}$$

Both expressions of D result in the same set, though the set builder notation is substantially more concise.

There are a few specific sets that we should mention by name. These sets will play a large role in much of our future work. We start with the *null* or *empty* set. As the name implies, this is a set with no elements.

DEFINITION 1.1. The Empty Set

$$\varnothing = \{\}$$

The set of *Natural Numbers* is a set consisting of all the positive whole numbers greater than zero.

DEFINITION 1.2. The Natural Numbers

$$\mathbb{N} = \{1, 2, 3, 4, \ldots\}$$

The three dots following the number 4 are called an *ellipsis*, and are used to signify that the presented pattern (in this case, adding one to the previous number) will continue indefinitely.

The set of *Integers* consists of all positive and negative whole numbers, and zero.

DEFINITION 1.3. The Integers

$$\mathbb{Z} = \{\ldots -3, -2, -1, 0, 1, 2, 3 \ldots\}$$

All numbers that can be written as a fraction make up the set of *Rational Numbers*. There is no clear pattern we should choose to represent the Rational Numbers (as we have previously done with the Natural Numbers and Integers), so we will instead use set builder notation.

DEFINITION 1.4. The Rational Numbers

$$\mathbb{Q} = \{x \mid x = \frac{a}{b} \text{ where } a, b \in \mathbb{Z} \text{ and } b \neq 0\}$$

In other words, a rational number is simply the result of dividing one Integer by another (provided that the denominator is not zero).

Next we consider the *Irrational Numbers*. The Irrational Numbers are those that cannot be represented as a fraction. They are decimal numbers that never terminate, and never repeat. A few examples of irrational numbers are $\sqrt{2}$, $\sqrt{3}$, and π.

DEFINITION 1.5. The Irrational Numbers

$$\{x \mid x \text{ is not a Rational Number}\}$$

You might notice that we did not assign a letter name to the set of Irrational Numbers (as we have previously done with the Natural Numbers, Integers, and Rational Numbers). There is no standard agreed upon representation for the set of Irrational Numbers. Whenever we need to make reference to it, we will simply say "The Irrational Numbers".

Finally, we arrive at the set of *Real Numbers*.

DEFINITION 1.6. The Real Numbers

$$\mathbb{R} = \{x \mid x \text{ is a Rational or Irrational Number}\}$$

The set of Real Numbers consists of all rational and irrational numbers, nothing more.

It is important to note that the representations of the Empty Set ($\varnothing$), the Natural Numbers ($\mathbb{N}$), Integers ($\mathbb{Z}$), Rational Numbers ($\mathbb{Q}$), and Real Numbers ($\mathbb{R}$) will only represent these specific sets. They will never be given any other meanings.

It will often be the case that one set S is part of another set T. In fact, if every element in S is also in T we say that S is a *subset* of T. There are two different notations we can use to illustrate S being a subset of T. The first is given by $S \subseteq T$ and would be read as "S is contained in T". The second is given by $T \supseteq S$ and would be read as "T contains S". In both cases, every element in the set S is also in the set T. As an example if $S = \{1, 2, 3\}$ and $T = \{1, 2, 3, 4, 5\}$ then S is a subset of T.

We finish this section with the idea of *set equality*. If we have two sets, S and T, then $S = T$ if S is a subset of T, $S \subseteq T$, and T is a subset of S, $T \subseteq S$. This relationship tells us that everything in S is also in T, and everything in T is also in S. Thus T and S have exactly the same elements and we consider them equal. If either S or T had even a single element that the other did not, they would not be considered equal.

CHAPTER 2

Function Theory

This chapter will develop the structure of functions. The ideas presented here extend far beyond the topic of precalculus. A good understanding of this material will simplify many of your future studies.

2.1. The Definition of a Function

To begin our exploration of functions, let us define two sets S and T such that

$$S = \{1, 2, 3, 4\} \quad \text{and} \quad T = \{a, b, c, d\}$$

We would like to pair all of the items in S with items in T. There are many, many different ways in which we could perform such an action. To avoid total anarchy, we will need some sort of structure that regulates these pairings. Our structure will go as follows; each item in S must pair with a unique item in T. For example, suppose we paired 1 with a. Then 1 cannot be paired with any other items in T.

How shall we write out the pairings between S and T? One common method is to use an ordered pair that follows the format

$$(s \in S, t \in T)$$

The pairing of 1 and a that we mentioned earlier would be $(1, a)$.

Aside from our single structure rule, there are no other restrictions on these pairings. There is no reason that 1 needs to pair with a, it could just as easily pair with any other element in T. Some examples of ways in which we could pair all of the elements in S with those in T, while following our rule, would be

$$(1, a), \ (2, b), \ (3, c), \ (4, d)$$

or

$$(1, d), \ (2, a), \ (3, b), \ (4, c)$$

or

2.1. THE DEFINITION OF A FUNCTION

$$(1,a),\ (2,a),\ (3,a),\ (4,a)$$

or

$$(1,b),\ (2,b),\ (3,c),\ (4,d)$$

Note that we are allowed to pair different elements of S with the same element in T (illustrated in the third and fourth examples) as this does not violate our single rule.

Some examples that do violate our rule would be

$$(1,a),\ (1,b),\ (2,b),\ (3,c),\ (4,d)$$

or

$$(1,d),\ (3,b),\ (4,c)$$

In the first example we have paired 1 with two different elements of T (a and b), and in the second we did not pair one of the elements of S (specifically, 2) with an element of T.

Another way in which we commonly refer to the pairings between the sets S and T is as a mapping from S to T. There is no change to the rule we laid down earlier, we are simply expanding on the way in which we think about connecting the items in S with the items in T. Coming back to our pairing of $(1,a)$ we would say that 1 maps to a.

We can draw a mapping of the pairings given by

$$(1,c),\ (2,b),\ (3,d),\ (4,a)$$

with a diagram showing the individual items in S being paired with the individual items in T.

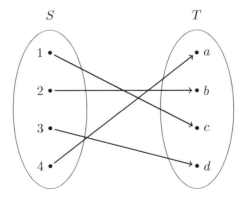

FIGURE 1. A mapping between S and T.

This diagram illustrates another important feature of our mapping, namely that we are starting in S and ending in T. When discussing a mapping between two sets, our starting set will be referred to as the *domain*, and the ending set is the *codomain*. In our example, the domain is S and the codomain is T.

DEFINITION 2.1. A *function* is a rule that pairs each element in the domain with a unique element in the codomain.

The terms function and mapping have the same meaning, and we will use them interchangeably.

Functions are generally named with lower case letters (though this is by no means required), and some extremely common choices are f, g, and h. The pairings shown in Figure 1 satisfy the definition of a function. Let us name this function f.

The notation we use to signify that f is a mapping from set S to set T is given by

$$f : S \to T$$

and we can observe how f pairs specific elements in the domain with those in the codomain by using the notation

$$f(s) = t \quad \text{where} \quad s \in S,\ t \in T$$

The notation $f(s)$ is read, "f of s" and represents the element in the codomain that is being paired with the element s from the domain. It is customary to refer to the element s as the *input* or *argument* and $f(s)$ as the *output* or *value*.

As an example, f pairs 1 with c (note that $1 \in S$ and $c \in T$) so we would write $f(1) = c$. The input to f was 1 and the output (or pairing) that f assigned was c. The remaining pairings in f are

$(2, b)$ an input of 2 gives an output of $f(2) = b$
$(3, d)$ an input of 3 gives an output of $f(3) = d$
$(4, a)$ an input of 4 gives an output of $f(4) = a$

Let us redraw Figure 1 using all of this new notation.

Figure 2 illustrates the idea that $f(1)$, $f(2)$, $f(3)$, and $f(4)$ are all elements of T (the codomain), while $1, 2, 3, 4$ are elements of S (the domain). The function itself, f, is the entire mapping of $S \to T$.

2.1. THE DEFINITION OF A FUNCTION

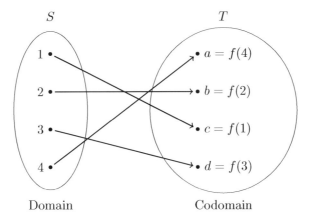

FIGURE 2. $f : S \to T$

Our function f is not the only possible function between S and T. There are many others. Another could be the function $g : S \to T$ that is given by the pairings

$$(1, b), \ (2, b), \ (3, a), \ (4, c)$$

or the function $h : S \to T$ given by

$$(1, a), \ (2, a), \ (3, 4), \ (4, 4)$$

Take a moment and verify that both g and h satisfy the definition of a function. Drawing h and g would give us the following.

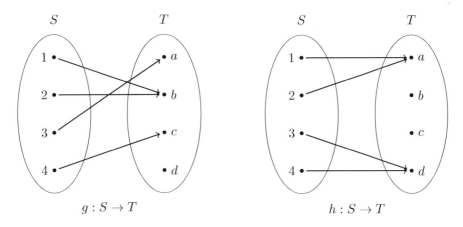

FIGURE 3. The mappings of g and h.

There are some stark differences between f and g or h. Both g and h map different elements of the domain to the same element in the codomain. Such pairing does not violate our definition of a function. We require each element in the domain to be paired with a unique element in the codomain. It does not matter if multiple elements in the domain map to the same element in the codomain. What matters is that each element in the domain maps to one distinct element in the codomain. Additionally, both g and h have elements in the codomain that are not paired with anything in the domain. Our definition of a function does not require that all elements of the codomain be paired with elements in the domain.

For any given function we can separate the codomain elements into two distinct subsets: those that are being paired by the function, and those that are not. The subset of the codomain that is being paired with elements in the domain is called the *range*. Consider again the functions g and h in Figure 3. The function g fails to pair the codomain element d with any element in the domain, so g has range $\{a, b, c\}$. Similarly, the function h fails to pair b and c with any elements in the domain, so h has range $\{a, d\}$.

Let us now consider some pairings that would not satisfy the definition of a function.

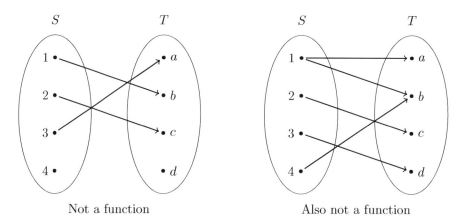

FIGURE 4. Pairings that do not satisfy the definition of a function.

The example on the left fails to map an element of the domain (the number 4) to the codomain. On the right, an element of the domain (the number 1) is being mapped to two different elements in the codomain. Both of these violate the definition of a function.

Exercises for Section 2.1.

(1) Let $A = \{5, 6, 7, 8\}$ and $B = \{1, 2, 3, 4\}$.

   (a) Write out the pairings of three different functions that have domain A and codomain B. Each of your three functions must have a different range.

   (b) Write out the pairings of three different functions that have domain B and codomain A. Each of your three functions must have a different range.

   (c) Using A as the domain, and B as the codomain, write out two different pairing schemes that violate the definition of a function. Explain why the pairing schemes you chose violate the definition.

(2) The function f consists of the following pairs:

$$(1, 4),\ (2, 3),\ (5, 0),\ (7, a),\ (c, 9)$$

   (a) What are the domain and range of f?

   (b) Evaluate the following

   (i) $f(5)$ (iv) $f(c)$

   (ii) $f(1)$ (v) $f(9)$

   (iii) $f(4)$ (vi) $f(6)$

(3) Let $h : S \to T$ be given by

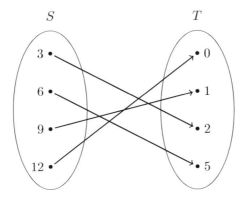

   (a) Write out all ordered pairs of the function h.

(b) What are the range and codomain of h?

(c) Express all $t \in T$ as $t = h(s)$ for the correct $s \in S$.

(4) How would you explain a function to someone who does not know the formal definition?

(5) Explain the difference between the codomain and the range of a function.

(6) Explain why the following statements are incorrect, and provide pairs between a domain and codomain that satisfy the given statement, but violate the definition of a function.

(a) A function is a rule that pairs each element in the domain with elements in the codomain.

(b) A function is a rule that pairs elements in the domain with unique elements in the codomain.

2.2. Equality, 1-1, and Onto

Now that we have established the definition of a function, and the idea that there are many possible functions between two sets, we must define what it will mean for two functions to be considered equal.

We have previously discussed the idea that a function consists of the entire mapping from the domain to the codomain, and we will use this notion to define the equality of two functions. Suppose we have two functions, f and g, each of which have domain S and codomain T. The only way in which we can consider f and g to be equal is if they make all of the same pairings between S and T.

DEFINITION 2.2. Given the functions f and g, each of which has domain S and codomain T, then $f = g$ if and only if $f(s) = g(s)$ for every $s \in S$.

It is important to note that function equality requires every pairing to be the same. If the two functions map a single element differently, they are not considered equal. Another subtle point is that we are requiring the codomain of of f and g to be the same. It is possible to relax this restriction and only require that the ranges of f and g be the same. We leave that idea for your future studies in set theory.

Next, let us reconsider two particular examples from the previous section. The functions f and g were defined by the following pairings.

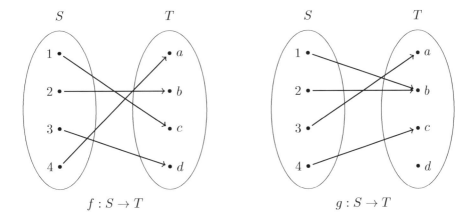

FIGURE 5. Revisiting the mappings of f and g.

We have already mentioned the idea that f and g seemed to operate in fundamentally different ways. The function g maps two different elements in the domain, 1 and 2, to the same element in the codomain, b. Additionally, g does not pair the element d of the codomain. Both of these behaviors play a large role in the world of functions. In fact, both of these behaviors have a distinct definition associated with them.

DEFINITION 2.3. If f is a function with domain S and codomain T, then f is *onto* or *surjective* if and only if for every $t \in T$ there is some $s \in S$ such that $t = f(s)$.

In other words, a function is considered onto if every element in the codomain is being paired with an element from the domain. When a function is onto, the codomain and the range are equal. In the examples from Figure 5, the function f is onto, but the function g is not (the element d is not being paired with anything in the domain).

DEFINITION 2.4. If f is a function with domain S and codomain T, then f is *one-to-one* or *injective* if and only if for every $s_1 \neq s_2$ in S, we have $f(s_1) \neq f(s_2)$ in T.

A function that is one-to-one maps each unique element in the domain to a unique element in the codomain. In other words, no two distinct elements in the domain are allowed to pair with the same element in the codomain. From the examples in Figure 5, the function f is 1-1, but the function g is not (the codomain elements 1 and 2 are both being paired with the domain element b). As mathematicians are nefariously efficient, we shorten the phrase one-to-one into 1-1. Whenever discussing functions, 1-1 signifies the one-to-one property that is defined above, not subtraction.

A function may be 1-1 and not onto, onto and not 1-1, both 1-1 and onto, or neither. Our function f from Figure 5 is both 1-1 and onto, while g is neither. Functions that are both 1-1 and onto will be of special interest to us.

DEFINITION 2.5. A *bijective* function is both 1-1 and onto.

If f is a bijection, it may also be referred to as a *1-1 correspondence*. An example of a bijective function is the *identity function*. We will name the identity function i (though we could just have easily named it any other letter), and define it as a map from some set S back to the same set S that is given by $i(s) = s$, which is to say that the input is the same as the output (or that every element is being paired with itself). Using our familiar set $S = \{1, 2, 3, 4\}$, the identity function would make the following pairings.

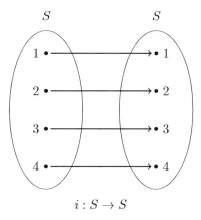

FIGURE 6. The identity function.

Exercises for Section 2.2.

(1) For the sets $S = \{a, b, c, d, e\}$, and $T = \{f, g, h, i\}$ the function $k : S \to T$ is defined by the following pairs

$$(a, g), \ (c, h), \ (b, i), \ (e, f), \ (d, h)$$

(a) Does k satisfy the properties of 1-1, onto, or bijective? Why or why not?

(b) For each $t \in T$, write $t = k(s)$ for the correct $s \in S$.

(2) For the sets $S = \{1, 2, 3, 4\}$, and $T = \{l, m, n, o, p\}$ the function $g : S \to T$ is defined by the following pairs

$$(2, m), \ (3, l), \ (1, p), \ (4, o)$$

(a) Does g satisfy the properties of 1-1, onto, or bijective? Why or why not?

(b) For each $t \in T$, write $t = g(s)$ for the correct $s \in S$.

(3) The functions f and g are defined by

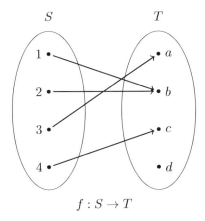
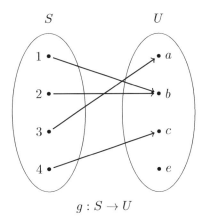

$f : S \to T$ \hspace{3cm} $g : S \to U$

(a) Are the codomains of f and g the same?

(b) What is the range of f?

(c) What is the range of g?

(d) Does $f = g$? Why or why not?

(4) Explain the following concepts.

(a) Function equality.

(b) The property of being onto.

(c) The property of being 1-1.

(d) The property of being bijective.

(5) Let $A = \{2, 4, 6, 8\}$ and $B = \{a, b, c, d\}$. The following exercises will ask you to define a function, $f : A \to B$, such that certain properties are met. If you are unable to define f with the required properties, explain why the problem cannot be completed, how the problem can be corrected, correct the problem, and then define the function.

(a) f is bijective.

(b) f is neither onto, nor 1-1.

(c) f is 1-1, but not onto.

(d) f is onto, but not 1-1.

(6) Recall the set of natural numbers, $\mathbb{N} = \{1, 2, 3, \ldots\}$. Suppose we define a set T, such that

$$T = \{t \mid t = 2 \cdot n \text{ where } n \in \mathbb{N}\}$$

or in other words, T is the set of even natural numbers, $T = \{2, 4, 6, \ldots\}$. We can define a function $f : \mathbb{N} \to T$ by

$$f(n) = 2 \cdot n$$

where $n \in \mathbb{N}$.

(a) Write out at least four different pairs of f.

(b) Do you think f is 1-1? Why or why not?

(c) Do you think f is onto? Why or why not?

(7) Suppose you know that $f : S \to T$ is bijective. What must that mean about the number of elements in S, and the number of elements in T? Consider (and explain) what your answer means with regard to the previous problem.

2.3. Function Composition

To begin this section, let us define three sets and two functions as follows.

$$S = \{1, 2, 3, 4\}$$
$$T = \{a, b, c, d\}$$
$$U = \{5, 6, 7\}$$

The function g is a bijection (1-1 and onto), while f onto, but not 1-1. Neither of those relationships have any bearing on what we are about to discuss, but it is good practice to notice. Note that set T serves as both the range of g, and the domain of f. Given this realization, it seems reasonable that we might redraw Figure 7 into the relationship shown in Figure 8.

The only difference between Figure 7 and Figure 8 is that the set T is only drawn once in Figure 8. Everything else is exactly the same. The new idea from Figure 8 is that when used together, these two functions provide us with a pairing between the items in S and the items in U.

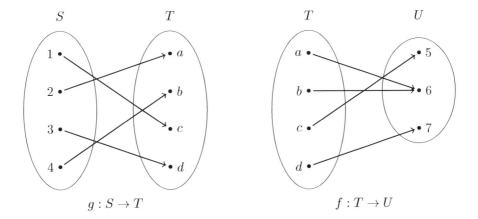

FIGURE 7. Two functions, f and g, based on the sets S, T, and U.

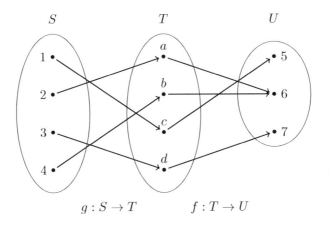

FIGURE 8. A redrawing of the functions f and g

As an example, we can see that the function g pairs the element 1 of S with the element c of T. The function f then pairs the element c of T with the element 5 of U. Through this two step process, we can consider 1 as being paired with 5. We are using the function g to map from $S \to T$, and then using the function f to continue the mapping from $T \to U$.

Let us consider this same pairing of $(1, 5)$ in a different fashion. We know that g gives us the pairing of $(1, c)$, thus

$$g(1) = c$$

Furthermore we have the pairing $(c, 5)$ from f, so
$$f(c) = 5$$

Suppose we substitute $g(1)$ in place of the c in the equation $f(c) = 5$. We would arrive with the notation
$$f(g(1)) = 5$$

which effectively suggests that we are using the output of the function g as the input for the function f. In fact, this is correct (and what is graphically illustrated in Figure 8).

DEFINITION 2.6. Given a function $g : S \to T$ and a function $f : T \to U$, the *composition* of f and g is a function with domain S and codomain U. The notation for this composition is written $f \circ g : S \to U$, where $(f \circ g)(s) = f(g(s))$ for every $s \in S$.

The composition $f \circ g$ is read "f of g". If we specify a particular element, s, that this function composition is acting on, we write this as $(f \circ g)(s)$ or $f(g(s))$. Both of those notations would be read as "f of g of s".

Don't let this notation (or any notation) intimidate you. The definition of function composition only formalized how the items in set S are paired with the items in set U (the definition is simply a formal explanation of Figure 8). The mechanics behind function composition are exactly as they were before the formal definition.

(1) Pick some element in set S (the domain of g). Let us call this element s.

(2) The function g will pair s with some element in set T, call this element $g(s)$.

(3) You're now at a specific element, $g(s)$, in set T. You know f is a function, with domain T and codomain U, so we can use $g(s)$ as input to f. This would be written as $f(g(s))$.

(4) We know that f will pair inputs with some element of U. Using function composition notation, we have $f(g(s)) = u$, where $u \in U$.

Exercises for Section 2.3.

(1) Suppose there are three sets, $A, B,$ and D, and these three sets are distinct (that is to say that $A \neq B \neq D$). If $f : A \to D$, $g : B \to A$, and both f and g are bijective, explain why the composition $g \circ f$ cannot possibly be a function, but the composition $f \circ g$ must be a function.

(2) The function g is defined by the following pairings

$$(1, b),\ (3, 5),\ (a, 9),\ (t, 1),\ (2, 2)$$

while the function f is defined by

$$(b, t),\ (5, 1),\ (2, 2),\ (1, a),\ (9, 3)$$

 (a) Is the composition $g \circ f$ a function? Why or why not?

 (b) Is the composition $f \circ g$ a function? Why or why not?

 (c) Does $f \circ g = g \circ f$? Why or why not?

 (d) Evaluate the following:

  (i) $f(g(1))$ (iv) $g(f(t))$ (vii) $g(g(3))$

  (ii) $f(g(a))$ (v) $g(f(5))$ (viii) $f(g(f(2)))$

  (iii) $f(g(2))$ (vi) $f(f(1))$ (ix) $g(f(g(5)))$

(3) Create three sets S, T and U, and two functions $g : S \to T$, $f : T \to U$, such that all of the following conditions are satisfied:

 - S, T, and U each contain 5 distinct elements.
 - $S \neq T \neq U$.
 - g is bijective.
 - f is not onto.
 - The composition $f \circ g$ is a function.

(4) Using your functions f and g from the previous problem, evaluate $f(g(s))$ for all $s \in S$.

(5) Suppose you have two functions, $f : T \to U$ and $g : S \to T$, and you know that g is not onto. Is it possible for the composition $f \circ g$ to be a function? Explain your answer.

2.4. Invertible Functions

Consider the function $f : S \to T$ shown in Figure 9 below.

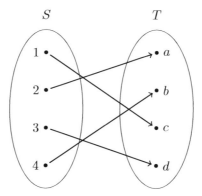

FIGURE 9. The bijection $f : S \to T$

Because f is bijective (1-1 and onto), it is possible for us to define a new function that maps from T to S utilizing the same pairing scheme as f, but in the opposite order. We refer to this function as f^{-1}, the *inverse* of f. When used on a function, the superscript of -1 signifies the inverse of the function. It is not an exponent. Visually, f^{-1} would be given by the following.

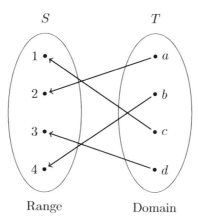

FIGURE 10. The bijection $f^{-1} : T \to S$

Note that the domain of f^{-1} is T and the range is S. The pairings of f and f^{-1} are shown together in the following table.

2.4. INVERTIBLE FUNCTIONS

f	f^{-1}
$(1,c)$	$(c,1)$
$(2,a)$	$(a,2)$
$(3,d)$	$(d,3)$
$(4,b)$	$(b,4)$

It is entirely reasonable to think of f^{-1} as "undoing" the mapping of f. As an example, we can see from both the table and Figure 9 that f maps 1 to c, and then f^{-1} maps c back to 1. Explore this relationship with the rest of pairings given by f and f^{-1}. In every case you will start with an element in S, map it to T with f, and then f^{-1} maps from T back to the very same element you started with in S. We can generalize this idea as follows.

Suppose f is a bijection with domain S and range T. Then for all $s \in S$ we have that $f^{-1}(f(s)) = s$ and that $f(f^{-1}(t)) = t$ for all $t \in T$.

As always, don't let the notation complicate the statement. We have only formalized the thought that the composition of a function and its inverse "undo" one another. The notation $f^{-1}(f(s)) = s$ is formally expressing the following steps.

(1) Pick an element in set S, call it s. The function f maps s to an element in set T, call this element $f(s)$.

(2) Use the element $f(s)$ as an input to f^{-1}.

(3) f^{-1} maps the element $f(s)$ back to s.

The notation $f(f^{-1}(t)) = t$ uses the exact same idea.

(1) Pick an element in set T, call it t. The function f^{-1} maps t to an element in set S, call this element $f^{-1}(t)$.

(2) Use the element $f^{-1}(t)$ as an input to f.

(3) f maps the element $f^{-1}(t)$ back to t.

Though it might not be immediately apparent, we have seen the idea behind the mappings of

$$f^{-1}(f(s)) = s \quad \text{and} \quad f(f^{-1}(t)) = t$$

before. At the time, we referred to it as i, the identity map. It is entirely accurate to say

$$f^{-1}(f(s)) = i(s) = s \quad \text{and} \quad f(f^{-1}(t)) = i(t) = t$$

In both compositions, the input element is mapped directly to itself.

Exercises for Section 2.4.

(1) The function h is defined by the following pairings

$$(a, c), \ (1, j), \ (5, z), \ (3, 0), \ (l, v)$$

Write out all pairings of h^{-1}.

(2) Come up with two sets, S and T, and a function, $f : S \to T$, such that f is not onto. Is $f^{-1} : T \to S$ a function? Why or why not?

(3) Come up with two sets, S and T, and a function, $f : S \to T$, such that f is not 1-1. Is $f^{-1} : T \to S$ a function? Why or why not?

(4) Based on the previous two problems, does it seem reasonable to conclude that only bijective functions will have inverses? Explain your answer.

(5) Complete the following table

x	$f(x)$	$f^{-1}(x)$
1		5
3	5	6
4		4
5		
	3	
9	6	1

Once the table has been completed, verify $f^{-1}(f(x)) = x$ for all elements in the domain of f, and $f(f^{-1}(x)) = x$ for all elements in the domain of f^{-1}.

CHAPTER 3

An Introduction to Real Functions

3.1. A Starting Point

With some general function theory behind us, we can now consider numeric examples of functions. As we are interested in the study of precalculus, almost all of the functions we discuss will have domains and ranges that are subsets of $\mathbb{R}$. To begin, let us consider the linear equation

$$y = \frac{2}{3}x + 2$$

This equation has slope $m = \frac{2}{3}$ and a y-intercept of 2. A graph would show the following.

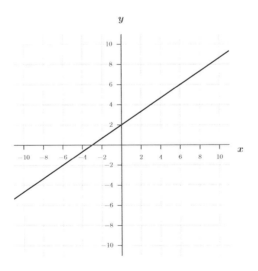

FIGURE 1. The linear equation $y = \frac{2}{3}x + 2$

In this linear equation we would say that y is the *dependent* variable, because the equation is solved for y, and the value we obtain for y depends on the value of x. Similarly, we would say that x is the *independent* variable because we are free to select a value we would like to use for x. As an example, if we let $x = 3$, then

$$y = \frac{2}{3}(3) + 2 = 4$$

We are able to determine that $y = 4$ because we chose $x = 3$. The output value for y required an input value for x.

We could just as easily make y the independent variable and x the dependent variable. To make this change, we would need to solve the equation for x. Doing so would yield

$$\frac{3}{2}(y - 2) = x$$

When we solve for x, the value we get for x depends on the value of y. We could now think of y as the input, and x as the output. While this train of thought is technically correct, and there is nothing wrong in treating x as the dependent variable, we will stick with the traditional slope-intercept form of a linear equation and leave things as

$$y = \frac{2}{3}x + 2$$

The solutions to our linear equation are ordered pairs of the form (x, y). The point $(3, 4)$ is one such solution. It is entirely reasonable to think that the x value of 3 is being paired with the y value of 4. It is not a coincidence that this "pairing" sounds similar to that of a function. In fact, with a slight change of notation, this linear equation will become our first numeric example of a function.

Many of our previous examples of functions involved pictures, where lines between elements in the domain and codomain illustrate the pairing scheme. While this is a perfectly valid approach to show the pairings of a function, it quickly falls into madness if the domain and codomain have a large number of elements. Imagine trying to draw a picture of a bijective function that had thousands of elements in the domain and codomain. Life is simply too short for such things. Rather than using a picture, we can use an equation to show how any given element in our domain will be paired with a unique element in our range.

3.1. A STARTING POINT

For a function named f we previously established the notation

$$f(\underbrace{\overbrace{\text{domain element}}^{\text{output}}}_{\text{input}}) = \underbrace{\text{range element}}_{\text{output}}$$

We will use this exact notation to write the linear equation

$$y = \frac{2}{3}x + 2$$

as a linear function. Treating the independent x value as the input, and the dependent y value as the output, we would have the following

$$f(x) = \frac{2}{3}x + 2$$

This notation tells us that we have a function named f where the domain element x is paired with a range element through the formula $\frac{2}{3}x + 2$. We would read $f(x)$ as "f of x." Let us graph this first linear function.

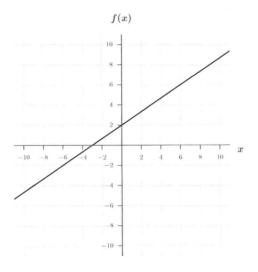

FIGURE 2. The linear function $f(x) = \frac{2}{3}x + 2$

The only visual difference between the linear equation shown in Figure 1 and the linear function in Figure 2 is the way in which we refer to the vertical axis. In Figure 1, the plotted line represents all of the pairings given by the equation

$$y = \frac{2}{3}x + 2$$

These pairings are points, each of which has the format (x, y), so we refer to axes as the x axis, and the y axis.

In the case of our function f shown in Figure 2, the plotted line represents all of the pairings given by the equation

$$f(x) = \frac{2}{3}x + 2$$

These pairings can also be interpreted as points, but they are of the form $(x, f(x))$. As such, it still makes sense to talk about the x axis, but we need to be careful when considering the vertical axis. The equation that gives us the pairs of the function f does not contain a y variable, yet it is still acceptable to refer to the vertical axis as the y axis, provided that we agree the y axis is the place where we will plot values of $f(x)$. Always keep in mind that when dealing with functions the vertical axis represents the range, in this case $f(x)$.

Before going any further we need to pause and verify that our linear function, f, satisfies the definition of a function. Namely, it must pair each domain element with a unique codomain element. But what are the domain and codomain of f?

During our explorations in precalculus, all of the functions we consider will be onto (there will not be elements in a codomain that are not paired with something in the domain). This simplification makes the range and codomain the same set, and will allow us to exclusively discuss the range from now on. We previously established that the x values are our domain values. The output value, $f(x)$ in the case of our linear function, will be a range value. This means that all of the ordered pairs making up the line shown in Figure 2 have the form

$$(\underbrace{\text{domain value}}_{x}, \underbrace{\text{range value}}_{f(x)})$$

Put another way, the domain of f is all of the values we are allowed to substitute for x in the equation

$$f(x) = \frac{2}{3}x + 2$$

Every time we substitute a value in for x we get a range value out for $f(x)$. Often times, there will be issues with certain real number substitutions for x. As an example, if we had the function

$$g(x) = \frac{1}{x}$$

we would not be able to substitute in $x = 0$. Doing so would result in division by 0 (thus causing the world to end). We will discuss such problems, and this specific example, in later sections. For now, our function $f(x) = \frac{2}{3}x + 2$ has no such issues and we are free to substitute any real number we would like for x. This means that the domain of f is $\mathbb{R}$.

The range of f consists of all output values, $f(x)$, that result from all possible input values of x. For any real number output you choose, there will be a corresponding real number input. As an example, suppose you choose $f(x) = 17$. For the number 17 to be in our range, we need some input value of x that results in an output value of 17.

$$17 = \frac{2}{3}x + 2$$

$$15 = \frac{2}{3}x$$

$$\frac{3}{2}(15) = x$$

$$\frac{45}{2} = x$$

So, an input value of $x = \frac{45}{2}$ will result in an output of $f(x) = 17$, and thus 17 is in the range of f. There is nothing special about our choice of 17, this argument can be applied to any real number (pick one and give it a try). This means that the range of f is $\mathbb{R}$.

With the domain and range determined, we can now verify that f is in fact a function. For f to satisfy the definition of a function it must pair each x input (domain value) with a unique $f(x)$ (range value). So, for any single value of x, there cannot be more than one unique value of $f(x)$. Visually, this would mean that no two points on the graph of f are on the same vertical line (if they were, they would have the same x value and different $f(x)$ values). As an example, if the points $(3, 4)$ and $(3, 5)$ were on our graph, then f would not be a function as 3 would not be paired with a unique range value. This idea is commonly referred to as the *vertical line test*. A visual inspection of Figure 2 and the knowledge that the graph will continue indefinitely with the same pattern (i.e. as a single straight line) verifies that f is in fact a function.

It is also possible to verify that f is a function algebraically. For any given value of x, the linear function f assigns a single unique $f(x)$ value (try substituting in some values for x to verify this idea for yourself). Thus f is a function. The vertical line test and algebraic verification of functions will always agree in their results. You should become comfortable utilizing both.

Exercises for Section 3.1.

(1) Explain the difference between dependent and independent variables. What are some real world examples of phenomena that are dependent on other things?

(2) Explain why any linear function that is non-vertical and non-horizontal must have a domain and range of $\mathbb{R}$. Your explanation should involve at least one sketch.

(3) Suppose you have a function named f such that $f(x) = 3x - 5$. Evaluate each of the following.

(a) $f(1)$

(b) $f(0)$

(c) $f(-3)$

(d) $f(\frac{1}{2})$

(e) $f(a)$

(f) $f(\pi)$

(4) Suppose there is a function named g such that $g(x) = \frac{2}{7}x - 1$. What domain values pair with the following range values?

(a) $g(x) = 1$

(b) $g(x) = 0$

(c) $g(x) = -2$

(d) $g(x) = \frac{3}{5}$

(e) $g(x) = a$

(f) $g(x) = \pi$

(5) Sketch examples of graphs that satisfy the vertical line test. Then, sketch examples of graphs that do not satisfy the vertical line test. For all graphs, explain why the vertical line test is able to verify whether or not a graph satisfies the definition of a function.

(6) Provide an example of a function that is not a graph.

3.2. Fundamental Functions

The purpose of this section is to introduce some of the most frequently occurring functions in precalculus, discuss them, and lay a foundation for much of what will be presented in the future. To begin, let us examine the *constant function*.

The general form of the constant function is given by

$$f(x) = a$$

where a is a real number. The constant function pairs all domain values with a. Put another way, this function consists of all points (x, a) where $x \in \mathbb{R}$. The domain of the constant function is all real numbers, and the range is $\{a\}$. Graphically, this function is a horizontal line. The example $f(x) = 3$ is shown below.

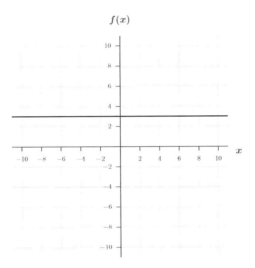

FIGURE 3. The constant function $f(x) = 3$

The general form of a *linear function* is given by

$$f(x) = mx + b$$

where m and b are real numbers, and m is not zero. This is also known as the slope-intercept form of a linear equation. The domain of the linear function is $\mathbb{R}$, and the range of the linear function is also $\mathbb{R}$. Graphically, this function is a straight line with non-zero slope. The example

$$f(x) = \frac{1}{2}x - 1$$

is shown below.

As we move from left to right on the x-axis, the corresponding values of $f(x)$ increase. Because increasing values of x result in increasing values of $f(x)$, we say that f is *increasing*. If increasing values of x resulted in decreasing values of $f(x)$, we would say that f was *decreasing*. Increasing and decreasing are properties that can apply to any function, not just linear ones.

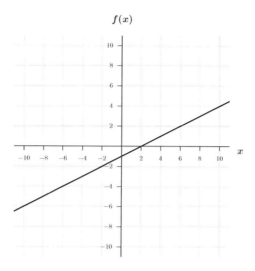

FIGURE 4. The linear function $f(x) = \frac{1}{2}x - 1$

The simplest general form of a *quadratic function* is given by

$$f(x) = x^2$$

The graph of a quadratic function is called a *parabola*.

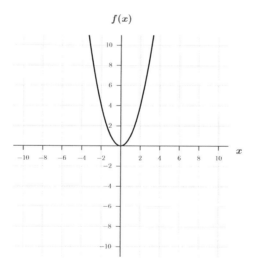

FIGURE 5. The quadratic function $f(x) = x^2$

The domain of this general quadratic function is ℝ, and the range is all real numbers greater than or equal to zero. In set notation, the range would be written

$$\{f(x) \mid f(x) \in \mathbb{R} \text{ and } f(x) \geq 0\}$$

This function is decreasing on the interval $-\infty < x \leq 0$ and increasing on the interval $0 \leq x < \infty$. The point where the function changes from decreasing to increasing, in this case $(0,0)$, is called a *turning point*. Any point where a function changes from increasing to decreasing, or decreasing to increasing, is a turning point. In the case of a parabola, we also refer to the turning point as the *vertex*.

Another property illustrated by this quadratic function is concavity. This function is *concave up* because it "opens upwards". If instead the parabola opened down (the vertex was the highest point instead of the lowest), we would call it *concave down*. While a substantially more robust definition of concave up and concave down awaits you in calculus, the idea of opening up and down is acceptable for the time being.

The simplest general form of a *cubic function* is given by

$$f(x) = x^3$$

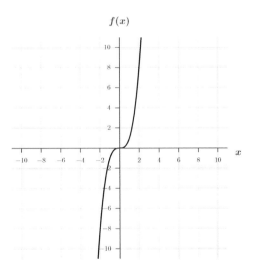

FIGURE 6. The cubic function $f(x) = x^3$

The domain and range of the cubic function are both ℝ, and the function is increasing. Additionally, the function is concave down on the interval $-\infty < x < 0$, and concave up on the interval $0 < x < \infty$. The point at which the concavity changes, and the function is defined, is called an *inflection point*. In the case of the fundamental cubic function the inflection point is $(0,0)$.

The general form of the *square root function* is given by

$$f(x) = \sqrt{x}$$

Before considering properties of the square root function, we should clarify the definition of a square root. The algebraic definition of a square root is given by

A number x is a square root of the number y if $x^2 = y$.

If we consider this definition in the context of a function, it allows two different input values to pair with the same output value. As a numeric example, this definition allows us to say that both 3 and -3 are square roots of the number nine as $(3)^2 = 9$ and $(-3)^2 = 9$. In function notation, we would have

$$f(9) = \sqrt{9} = 3 \quad \text{and} \quad f(9) = \sqrt{9} = -3$$

which violates the definition of a function. The algebraic definition of a square root requires an addendum if we are to make the square root a function. The addendum is that we define the square root to mean only the positive answer. This is often referred to as the *principal root*. The positive and negative roots are respectively given by

$$\sqrt{9} = 3 \quad \text{and} \quad -\sqrt{9} = -3$$

and now the square root satisfies the definition of a function.

As the square root of a negative number is not a real number, the domain of the square root function is limited to all real numbers greater than or equal to zero

$$\{x \mid x \in \mathbb{R} \text{ and } x \geq 0\}$$

The square root of a positive number is a positive number, so the range must also be greater than or equal to zero. Furthermore, any real number greater than or equal to zero will be in the range. To verify this fact, choose any real number you desire to be in the range. The square of that number is in the domain (the domain is all real numbers greater than zero), so the square root of that number is in the range. Thus the range is the same set as the domain

$$\{f(x) \mid f(x) \in \mathbb{R} \text{ and } f(x) \geq 0\}$$

The square root function is increasing, and concave down.

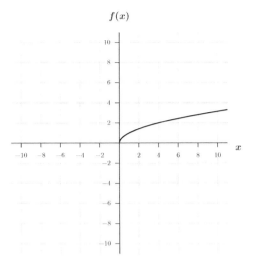

FIGURE 7. The square root function $f(x) = \sqrt{x}$

The general form of the *reciprocal function* is given by

$$f(x) = \frac{1}{x}$$

We need to spend a moment considering the domain of the reciprocal function. Specifically, we need to state that zero cannot be in the domain. If zero were in the domain, then $f(0) = \frac{1}{0}$ would be in the range (and as we previously mentioned, this causes the world to end). This means that the domain of the reciprocal function is all real numbers other than zero

$$\{x \mid x \in \mathbb{R} \text{ and } x \neq 0\}$$

The range of the reciprocal function has a similar limitation. Pick any real number you desire to be in the range, suppose we call it a. In order for a to be in the range, there needs to be an element in the domain that maps to a. We determine if this mapping exists by setting $f(x) = a$ and solving for x, the domain value that would pair with a.

$$f(x) = \frac{1}{x}$$

$$a = \frac{1}{x}$$

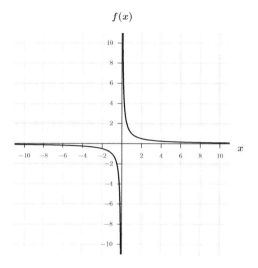

FIGURE 8. The reciprocal function $f(x) = \frac{1}{x}$

$$ax = 1$$

$$x = \frac{1}{a}$$

So, the domain value that pairs with a is $\frac{1}{a}$. The only value for a that presents us with any trouble is $a = 0$ (we would again be dividing by zero), so zero is not in the range of the reciprocal function. All other real numbers are in the range, making it the same set as the domain

$$\{f(x) \mid f(x) \in \mathbb{R} \text{ and } f(x) \neq 0\}$$

The reciprocal function is our first example of a *discontinuous* function. So far, all other functions that we have viewed have been *continuous*. A complete definition of continuous and discontinuous requires topics that are generally reserved for calculus, but for now we can say that a function is continuous if its domain consists of more than a single point, and it is possible for you to draw the graph of the function without your pen leaving the page. If you must pick your pen up and place it back down, then the function is discontinuous. The reciprocal function is discontinuous at $x = 0$. At an x value of 0 you must lift your pen, and place it back down on the other side of $x = 0$ to continue drawing.

The reciprocal function is also our first example of a function that has an *asymptote*. An asymptote is a line (or curve) that a function will get arbitrarily close to. The reciprocal function has two asymptotes, the first is the vertical line $x = 0$. The reciprocal function will get as close as we would like to this vertical line, but we will never be able to reach it.

To illustrate this idea, suppose we wanted to know where the reciprocal function was a horizontal distance of 1 away from the vertical asymptote $x = 0$. There are two points when the reciprocal function is a horizontal distance of 1 away from the vertical line $x = 0$. The x coordinates of these two points are $x = 1$ and $x = -1$. To solve for the vertical $f(x)$ coordinate, we need only to plug these domain values into f.

$$f(1) = \frac{1}{1} = 1 \quad \text{or} \quad f(-1) = \frac{1}{-1} = -1$$

Both of these points, $(1, 1)$ and $(-1, -1)$ respectively, are a horizontal distance of 1 away from the vertical line $x = 0$. We could repeat this process for any horizontal distance we wish. Suppose we wanted to know where the reciprocal function is a horizontal distance of $\frac{1}{2}$ away from the vertical asymptote $x = 0$.

$$f\left(\frac{1}{2}\right) = \frac{1}{\frac{1}{2}} = 2 \quad \text{or} \quad f\left(-\frac{1}{2}\right) = \frac{1}{-\frac{1}{2}} = -2$$

Both of these points, $(\frac{1}{2}, 2)$ and $(-\frac{1}{2}, -2)$, are a horizontal distance of $\frac{1}{2}$ away from the vertical line $x = 0$. We can get as close to the line $x = 0$ as we like, but we can never actually get there (doing so would require an input of 0, and 0 is not in the domain of the reciprocal function).

In addition to the vertical asymptote at $x = 0$, the reciprocal function has a horizontal asymptote at a vertical value of 0. We write the equation of the horizontal asymptote as

$$y = 0$$

The reciprocal function will get as close as we would like to the horizontal line $y = 0$, but it will never actually get there (doing so would require division by 0). The arguments and ideas concerning distance from this horizontal asymptote are analogous to those we gave for the vertical asymptote, the only difference being that we would now be interested in the vertical distance from the horizontal line $x = 0$.

Our final fundamental function (for now) will be the *absolute value function*. If you have previously worked with the absolute value, you may have heard it defined as "the distance from zero". That definition is correct, but it does not provide any kind of an algebraic structure we can manipulate. If we wish to perform algebra with the absolute value, we require an algebraic definition.

The general form of the absolute value function is given by

$$f(x) = \begin{cases} x & \text{if } x \geq 0 \\ -x & \text{if } x < 0 \end{cases}$$

This definition does not contradict the distance interpretation of the absolute value, and the absolute value still operates exactly as you remember. Consider the following examples.

$$|5| = 5 \quad \text{and} \quad |-5| = -(-5) = 5$$

The absolute value is our first example of a *piecewise* function. A piecewise function is a function that is defined in pieces, based on the domain value that is used as input. The absolute value function is defined in two pieces. If you are trying to pair a domain value that is greater than or equal to zero, then the pairing is given by $f(x) = x$. If you are trying to pair a domain value that is less than zero, then the pairing is given by $f(x) = -x$. To illustrate this idea, let us compare the pairings of 3 and -3

$$f(3) = 3 \quad \text{and} \quad f(-3) = -(-3) = 3$$

which would correspond to the points $(3, 3)$ and $(-3, 3)$ respectively. The graph of the absolute value function is shown in Figure 9.

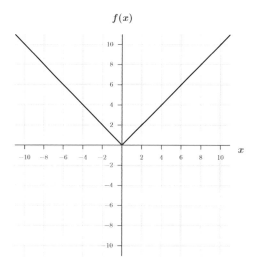

FIGURE 9. The absolute value function $f(x) = |x|$

The domain of the absolute value function is $\mathbb{R}$. As illustrated by the previous example, it is impossible for the range to contain a negative number, thus the range is

$$\{f(x) \mid f(x) \in \mathbb{R} \text{ and } f(x) \geq 0\}$$

Exercises for Section 3.2.

(1) Let the functions f, and g be defined by

$$f(x) = \frac{2}{5}x + 6 \quad \text{and} \quad g(x) = x^2$$

Evaluate each of the following:

(a) $f(3)$ (b) $g(5)$ (c) $f(\pi)$ (d) $g(x+3)$

(2) Explain why any version of the equation

$$x = a$$

where $a \in \mathbb{R}$, can never be a function.

(3) Explain the relationship between the slope of a linear function and whether the function is increasing or decreasing. Can a linear function both increase and decrease? Why or why not?

(4) Given the fundamental square root function, $f(x) = \sqrt{x}$, determine the domain values that pair with the following range values. Classify each of your answers as rational or irrational.

(a) $f(x) = 9$ (b) $f(x) = \pi$ (c) $f(x) = 2$ (d) $f(x) = 169$

(5) Determine whether or not each of the fundamental functions are 1-1. Your answers must each include a sketch that clearly illustrates your reasoning.

(6) Graph the following piecewise functions, and state their respective domains. Think carefully about what must happen for the strict inequalities.

(a)
$$f(x) = \begin{cases} x & \text{if } x \geq 7 \\ -x & \text{if } 3 \leq x < 7 \\ |x| & \text{if } -4 < x < 3 \end{cases}$$

(b)
$$g(x) = \begin{cases} \sqrt{x} & \text{if } x > 9 \\ 5 & \text{if } 3 < x \leq 9 \\ x^2 & \text{if } 2 < x < 3 \\ \frac{1}{x} & \text{if } x \leq 2 \end{cases}$$

(7) Let h be a linear function. Suppose that $h(0) = 3$ and $h(-3) = 5$. What is the linear equation for $h(x)$?

(8) Let f be a linear function. Suppose that $f(0) = 3$ and $f(x) - f(x+1) = -2$ for all values of x. What is the linear equation for $f(x)$?

(9) Let g be the linear function given by $g(x) = 2x - 1$.

  (a) What would $g(x+1) - g(x)$ be for all values of x?
  (b) What would $g(x+2) - g(x)$ be for all values of x?
  (c) What would $g(x+3) - g(x)$ be for all values of x?

3.3. The Domain of a Function

There are some very common (and very useful) exercises we can employ to help us gain familiarity with real functions. Suppose we have a function f whose pairings are given by an equation

$$f(x) = \text{some expression involving } x$$

We have previously discussed the idea that the x variable represents a domain (or input) value, and that $f(x)$ represents the pairing range value. Given that relationship, it seems reasonable to wonder exactly which real numbers might make up the domain

3.3. THE DOMAIN OF A FUNCTION

and range of f. This is not always an easy question to answer. For now, let us start by exploring the domain of a function through some examples.

Consider the function

$$f(x) = \frac{2x - 5}{x - 3}$$

We know that all of the values in the domain of f will be paired with their range value through the above equation. Furthermore, our experience with the reciprocal function tells us that we cannot use any input (x value) that will result in division by zero. In the case of f, that x value is 3.

$$f(3) = \frac{2(3) - 5}{3 - 3} = \frac{6 - 5}{0}$$

Thus, f cannot pair an input value of 3. This means that 3 is not in the domain of f. There are no other real numbers that f is unable to pair, so the domain of f would be

$$\{x \mid x \in \mathbb{R} \text{ and } x \neq 3\}$$

This example of an input value for x resulting in division by zero is called an *asymptotic discontinuity*. We have previously seen two types of asymptotes, vertical and horizontal. The zeros of a denominator are vertical asymptotes (they may also be holes, but we will cover that topic in a later section). The asymptote is vertical because there can be no output (vertical) value for an input (horizontal) value that results in the denominator of a fraction being equal to zero.

Sometimes the zeros of a denominator are not immediately apparent and we need to perform some algebra to identify them. The function g that is given by the equation

$$g(x) = \frac{9 - x}{x^2 - 4x - 5} = \frac{9 - x}{(x + 1)(x - 5)}$$

cannot be evaluated for $x = -1$ or $x = 5$ (both result in division by zero). Thus the domain of g would be

$$\{x \mid x \in \mathbb{R} \text{ and } x \neq -1 \text{ and } x \neq 5\}$$

and there are vertical asymptotes at $x = -1$ and $x = 5$.

How might something like a square root influence a function's domain? Let us define the function h as

$$h(x) = \sqrt{8 - 5x}$$

What is the domain of h? If the radicand (what's under the square root) is less than zero, we will take the square root of a negative number. While this is possible (there are entire branches of mathematics devoted to this idea), the square root of a negative real number is not a real number. In order for us to work only with real numbers, we need the radicand to be at least 0. Thus, we require that $8 - 5x \geq 0$. Solving this linear inequality will provide us with the domain of h.

$$8 - 5x \geq 0$$

$$-5x \geq -8$$

$$x \leq \frac{8}{5}$$

In order for the radicand to be at least zero, we must have $x \leq \frac{8}{5}$. The domain of h is then

$$\left\{ x \mid x \in \mathbb{R} \text{ and } x \leq \frac{8}{5} \right\}$$

In short, whenever a function involves a square root, we need the radicand to be at least zero. The process of determining where a radicand is at least zero is straightforward, provided that the radicand is linear (the radicand from the function h was $8 - 5x$, which is linear). As was illustrated, setup a linear inequality where the radicand is greater than or equal to zero, and solve.

Nonlinear radicands require a slightly more involved approach. Given the function

$$f(x) = \sqrt{x^2 - 2x - 3}$$

we still require the radicand to be at least zero, but how exactly shall we determine where $x^2 - 2x - 3 \geq 0$? Let us consider $x^2 - 2x - 3$ as a function in its own right, suppose we call it g. The question we are trying to answer is what input values of x result in an output that is zero or greater for

$$g(x) = x^2 - 2x - 3$$

We begin exploring the answer to this question by graphing g. As shown in Figure 10, the x values for which $g(x)$ is greater than or equal to zero are $x \leq -1$ and $x \geq 3$. This idea can be illustrated visually if we redraw Figure 10 with a domain of $\{x \mid x \in \mathbb{R} \text{ and } x \leq -1 \text{ or } x \geq 3\}$.

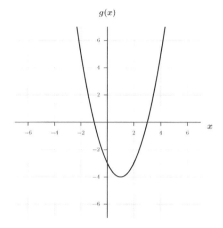

FIGURE 10.
$g(x) = x^2 - 2x - 3$

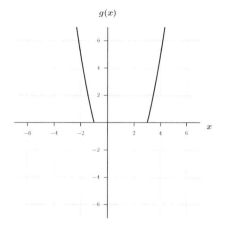

FIGURE 11. g with domain $x \leq -1$ or $x \geq 3$

Thus, the domain of the function

$$f(x) = \sqrt{x^2 - 2x - 3}$$

is $\{x \mid x \in \mathbb{R} \text{ and } x \leq -1 \text{ or } x \geq 3\}$ as those are the x values for which the radicand, $x^2 - 2x - 3$, is greater than or equal to zero.

While graphing is a helpful tool for determining the sign of a function, we will need a more exact algebraic method. Recall the previously mentioned idea of continuity (a function is continuous if you can draw it without removing your pen from the page). If a function is positive (vertical output values are above the x-axis) and continuous, there is no way for that function to become negative without crossing the x-axis. Similarly, if a function is negative and continuous, there is no way for that function to become positive without crossing the x-axis. This means that the zeros of a continuous function (the input values that result in an output value of zero, i.e. where the function crosses the x-axis) represent all possible locations that the function can change its sign.

In Figure 10, the function g changed sign from positive to negative after crossing the zero at $x = -1$. It remained negative on the interval $-1 < x < 3$, and then changed sign again after crossing the zero at $x = 3$. Because g is continuous, there are no other possible locations where it could change the sign of its output. We can use this fact to rapidly determine where a function is positive or negative by creating a *sign chart*.

To create a sign chart, first write the function as a product of factors. Our function g would be

$$g(x) = x^2 - 2x - 3 = (x-3)(x+1)$$

Write the factors down the left side of a table, with the final factored form of the function last. Next, write the zeros of the function, from least to greatest, across the top of the table. Leave a space to the left and right of each zero.

		-1		3	
$x+1$					
$x-3$					
$(x+1)(x-3)$					

Starting with the factor $x+1$, let x be any real number to the left (less than) the zero of -1. Regardless of your choice, $x+1$ will be less than zero, so the sign of $x+1$ is negative. We fill this result into the sign chart as follows.

		-1		3	
$x+1$	$-$				
$x-3$					
$(x+1)(x-3)$					

Continuing down the first column, $x-3$ is also less than zero for any value of x that is to the left of -1. The last row of the first column is the factored form of our function, in this case the product of these two factors. Both of the factors are negative, and two negative numbers multiplied together result in a positive number.

		-1		3	
$x+1$	$-$				
$x-3$	$-$				
$(x+1)(x-3)$	$+$				

Continuing with this idea, we can fill in the rest of the sign chart.

		-1		3	
$x+1$	$-$	0	$+$	$+$	$+$
$x-3$	$-$	$-$	$-$	0	$+$
$(x+1)(x-3)$	$+$	0	$-$	0	$+$

The last row of our sign chart tells us that $g(x) = x^2 - 2x - 3 = (x-3)(x+1)$ is greater than or equal to zero for $x \leq -1$ or $x \geq 3$, which means that our function

$$f(x) = \sqrt{x^2 - 2x - 3}$$

has a domain of $\{x \mid x \in \mathbb{R} \text{ and } x \leq -1 \text{ or } x \geq 3\}$. This is the exact result we obtained earlier.

Sign charts can also be used to determine the sign of a discontinuous function. Consider the function

$$f(x) = \frac{2x-3}{x-5}$$

A fraction is only ever equal to zero when the numerator is zero. Setting the numerator equal to zero and solving gives us

$$2x - 3 = 0$$

$$2x = 3$$

$$x = \frac{3}{2}$$

Thus f has one zero, $x = \frac{3}{2}$. Additionally, f has a vertical asymptote at $x = 5$. Discontinuous functions can change sign across discontinuities. This means that a sign chart for a discontinuous function must consider not only the zeros, but also the discontinuities. The sign chart for f would be

		$\frac{3}{2}$		5	
$2x-3$	$-$	0	$+$	$+$	$+$
$x-5$	$-$	$-$	$-$	0	$+$
$\frac{2x-3}{x-5}$	$+$	0	$-$	UN	$+$

where the UN is shorthand for "undefined". Note that f changed sign across the zero at $x = \frac{3}{2}$ and the vertical asymptote $x = 5$.

At this point a somewhat devious math instructor might ask "So, what is the domain of f?" While this example illustrated how to construct a sign chart for a discontinuous function, the function we were considering does not require a sign chart to determine its domain. Fractions can be both positive and negative, the only thing we need to worry about is where the denominator is equal to zero. The domain of f is $\{x \mid x \in \mathbb{R} \text{ and } x \neq 5\}$.

Exercises for Section 3.3.

(1) Determine the domain of each of the following functions. Construct sign charts where appropriate.

(a) $f(x) = \sqrt{7 - 9x}$

(b) $g(x) = (x + 4)^3 - 1$

(c) $h(x) = \sqrt{x^2 - 3x + 18}$

(d) $j(x) = 6$

(e) $k(x) = \dfrac{\sqrt{8 - x}}{x + 8}$

(f) $l(x) = \dfrac{x + 4}{2x - 5}$

(g) $m(x) = |x + 11|$

(h) $n(x) = \sqrt{x + 1} \cdot \sqrt{x - 5}$

(i) $p(x) = \dfrac{1}{\sqrt{x^2 + x - 6}}$

(j) $q(x) = \sqrt{x^3 + 5x^2}$

(k) $r(x) = \sqrt{\dfrac{x^2 - 4}{x - 9}}$

(l) $s(x) = \dfrac{2x^2 + 5x}{x + 10}$

(2) Consider the function

$$f(x) = \dfrac{6x^2 - 5}{x + 2}$$

Using algebra, determine which of the following values are in the range of the function f. Is f 1-1?

(a) 2   (b) -10 (c) 0   (d) -50

(3) The given figure illustrates all pairings of the function f. What are the approximate domain and range of f? Is f 1-1?

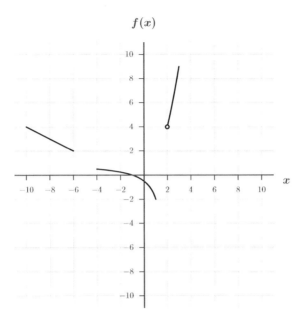

3.4. Even, Odd and Equality

This section will cover two new definitions, and revisit our earlier definition of function equality. Let us begin with the new definitions.

DEFINITION 3.1. Even Function

A function f is *even* if and only if $f(x) = f(-x)$ for all x in the domain of f.

The equation $f(x) = f(-x)$ is saying that the output (vertical value) for an input of x is the same as the output for an input of $-x$.

$$\overbrace{f(x)}^{\text{Output for an input of } x} = \underbrace{f(-x)}_{\text{Output for an input of } -x}$$

Though it wasn't mentioned at the time, we have already seen examples of even functions.

Consider the fundamental function $f(x) = x^2$. Let us compare the output values for input values of 2 and -2.

$$f(2) = 2^2 = 4 \qquad \text{and} \qquad f(-2) = (-2)^2 = 4$$

So, $f(2) = f(-2)$, and we have our first numeric example of an even function. Note that this example does not prove that f is an even function. In order for f to be even we must have the relationship $f(x) = f(-x)$ for all x in the domain of f. We have only shown that $f(2) = f(-2)$. Proving that f is even would require us to verify this relationship for all real numbers if the domain of f. Given that the domain of f is $\{x \mid x \in \mathbb{R}\}$, we can't attempt to show this fact one real number at a time (we would be here forever, literally). Instead, we can verify things algebraically.

$$f(-x) = (-x)^2 = x^2 = f(x)$$

Verifying the definition with a generic x proves that $f(x) = x^2$ for any specific value of x we might choose. Thus, f is an even function.

Even functions are symmetric about the vertical axis. This fact follows directly from the definition. Imagine you are graphing an even function. For an input of x, you would start at the origin and move a horizontal distance of x. From there, you would move a vertical distance of $f(x)$. For the input $-x$ you would move the same distance you moved as the x input, but in the opposite direction. Then, you would move a vertical distance of $f(-x) = f(x)$, because the function is even.

The points $(2, 4)$ and $(-2, 4)$ on the graph of the function $f(x) = x^2$. To plot the point $(2, 4)$ you move 2 horizontally and then 4 vertically. The point $(-2, 4)$ has you moving the same horizontal distance, 2, but in the opposite direction (the negative sign). Then, you move the same vertical distance in the same direction, 4. This pattern will continue for all values in the domain of f, making the graph of f symmetric about the vertical axis.

DEFINITION 3.2. Odd Function

A function f is *odd* if and only if $f(-x) = -f(x)$ for all x in the domain of f.

Again, don't make the definition more complicated than it is. The output for an input of $-x$ is the same as making the output for an input of x, negative.

$$\underbrace{f(-x)}_{\text{Output for an input of } -x} = \underbrace{-f(x)}_{\text{Negative of the output for an input of } x}$$

We have also already worked with examples of odd functions.

3.4. EVEN, ODD AND EQUALITY

Consider the fundamental function $f(x) = x^3$. Let us compare the output value for inputs of -3 and 3.

$$f(-3) = (-3)^3 = -27 \quad \text{and} \quad f(3) = 3^3 = 27$$

Note that we do not have $f(-3) = f(3)$, this is not an even function. However, if we make $f(3)$ negative

$$-f(3) = -(3^3) = -27$$

then we have $f(-3) = -f(3)$. This is our first numeric example of an odd function. Again, this does not prove that f is an odd function. We would need to verify this relationship for all x in the domain of f. Doing so one real number at a time would take forever, so we verify the relationship algebraically.

$$f(-x) = (-x)^3 = -x^3 \quad \text{and} \quad -f(x) = -(x^3) = -x^3$$

Thus $f(-x) = -f(x)$ and we have proven that $f(x) = x^3$ is an odd function.

Odd functions are symmetric about the origin. This fact follows directly from the definition. Imagine graphing the points $(x, f(x))$ and $(-x, f(-x))$. The inputs of x and $-x$ are the same horizontal distance from the origin, but in opposite directions. Because the function is odd, the vertical values $f(-x)$ and $f(x)$ are the same distance from the x-axis, but in opposite directions. This is why the function is symmetric about the origin.

The points $(3, 27)$ and $(-3, -27)$ are on the graph of the function $f(x) = x^3$. Both points have a horizontal distance of 3 away from the origin, but in opposite directions. From there, both points have a vertical distance of 27 away from the x-axis, but in opposite directions. This pattern will continue for all points on the graph of f, resulting in symmetry about the origin.

With those two definitions completed, we can now revisit the idea of function equality. Two functions are equal if and only if they make all of the same pairings between the domain and codomain (but recall that all of the functions we are working with are onto, so we can just consider the range). That definition is correct, and not in need of any modification. What we do need, are some examples.

Consider the functions

$$f(x) = \frac{x+3}{(x+3)(x-2)} \quad \text{and} \quad g(x) = \frac{1}{x-2}$$

it is a simple matter to show that

$$\frac{\cancel{x+3}}{\cancel{(x+3)}(x-2)} = \frac{1}{x-2}$$

and we are left wondering, does this allow us to say that the functions f and g are equal? In fact, these functions are not equal. A function is defined by exactly what is given, not by how it may be simplified. Our ability to algebraically simplify the formula that tells us how to pair the domain of f with the range of f does not change the domain of f. The domain of f is $\{x \mid x \in \mathbb{R} \text{ and } x \neq -3 \text{ and } x \neq 2\}$, while the domain of g is $\{x \mid x \in \mathbb{R} \text{ and } x \neq 2\}$. The real number -3 is not in the domain of f, but it is in the domain of g.

$$f(-3) \text{ is undefined} \quad \text{and} \quad g(-3) = \frac{1}{-3-2} = -\frac{1}{5}$$

This means that $f \neq g$ as they do not pair the domain element -3 in the same way.

As another example, let f and g be given by

$$f(x) = \frac{\sqrt{x-5}}{\sqrt{x-5}} \quad \text{and} \quad g(x) = \frac{\sqrt{x-7}}{\sqrt{x-7}}$$

The domain of f is $\{x \mid x \in \mathbb{R} \text{ and } x > 5\}$, while the domain of g is $\{x \mid x \in \mathbb{R} \text{ and } x > 7\}$. This immediately tells us that $f \neq g$ (f can pair all real numbers between 5 and 7, while g cannot).

Suppose that we specify the domain of f to be $\{x \mid x \in \mathbb{R} \text{ and } x > 7\}$. This specification prevents f from making all of the pairings that f and g previously disagreed upon. Both of these functions are now defined on the same domain, and pairing all elements of the domain and range in exactly the same way (specifically, they both pair all of their domain elements with the number 1). Now we can say that $f = g$.

Exercises for Section 3.4.

(1) This section contained simple proofs that $f(x) = x^2$ is an even function, while $f(x) = x^3$ is an odd function. Using those as a guide, prove which (if any) of the remaining fundamental functions are even, and which are odd.

(2) Can a function be both even and odd? Explain your reasoning.

(3) Suppose f is a function whose graph is symmetric about the line $x = 4$. If f makes the pairing $(1, 3)$, what other pairing must f make?

(4) There are two functions, f and g. The function g makes its pairings according to the relationship

$$g(x) = f(x) + f(-x)$$

(a) Suppose you know that f is an even function. What can you determine about the symmetry of g?

(b) Suppose you know nothing about f. Can you determine anything about the symmetry of g?

(5) There are two functions, f and h. The function h makes its pairings according to the relationship

$$h(x) = f(x) - f(-x)$$

(a) Suppose you know that f is an odd function. What can you determine about the symmetry of h?

(b) Suppose you know nothing about f. Can you determine anything about the symmetry of h?

(6) Classify the following functions as having even symmetry, odd symmetry, or neither. Justify your choice.

x	$f(x)$
3	-12
-1	-4
8	-7
1	-4
-3	-12
-8	-7

x	$g(x)$
a	11
5	b
9	$-c$
c	9
-11	a
$-b$	5

x	$h(x)$
2	-15
-6	-23
10	0
-10	0
6	23
-2	15

(7) For each part of this problem, begin by sketching the graph of $f(x) = \sqrt{x}$. Then, modify the graph as instructed.

   (a) Modify the graph so that it has even symmetry.

   (b) Modify the graph so that it has odd symmetry.

   (c) Modify the graph so that it has symmetry across $x = 2$.

(8) Determine if the functions f and g are equal. Justify your answers.

   (a)
   $$f(x) = \frac{\sqrt{x-3}}{\sqrt{x+4}} \qquad g(x) = \sqrt{\frac{x-3}{x+4}}$$

   (b)
   $$f(x) = \left(\sqrt{x+1}\right)^2 \qquad g(x) = x+1$$

   (c)
   $$f(x) = \sqrt{(x+1)^2} \qquad g(x) = x+1$$

   (d) Let $a, b \in \mathbb{R}$ such that $0 < b < a$
   $$f(x) = \left(\sqrt{\frac{x+a}{x+b}}\right)^2 \qquad g(x) = \frac{x+a}{x+b}$$

   Assume the domains of f and g are limited to $\{x \mid x \in \mathbb{R} \text{ and } x > -b\}$.

3.5. Vertical Transformations

The next several sections are dedicated to function transformations. For our purposes, a function transformation will be the process of taking the graph of an arbitrary function and shifting, stretching, compressing, or rotating it about an axis. These movements can be separated into two distinct groups; those that happen horizontally and those that happen vertically. We will discuss the vertical first.

Let us begin with a function that is familiar, $f(x) = x^2$. We know that f will contain the following pairings.

3.5. VERTICAL TRANSFORMATIONS

x	$f(x)$
-2	4
0	0
2	4

These three points do not represent all possible pairings of f; they are random choices. All possible pairings of f can be represented by points on a graph, and that graph is a parabola whose vertex is the origin. Suppose we want to shift this parabola up 2 units. A vertical shift does not involve any horizontal change. All of the x values in the previous table will remain the same, while we add 2 to the vertical values.

x	$f(x) + 2$
-2	6
0	2
2	6

Remember, $f(x) = x^2$, so the function $f(x) + 2 = x^2 + 2$. To avoid confusion, let us call this new function g. The new function $g(x) = x^2 + 2$ is the function f shifted up 2 units, $g(x) = f(x) + 2$.

We can just as easily shift a function down. Suppose now we let $f(x) = x^3$. Some of the pairings of f would be

x	$f(x)$
-2	-8
0	0
2	8

If we wanted to shift f down 5 units, we would need to subtract 5 from all of the vertical values (again leaving the horizontal values alone). This would give us

x	$f(x) - 5$
-2	-13
0	-5
2	3

where $f(x) - 5 = x^3 - 5$. For the sake of clarity, let's give this new function the name g. The function $g(x) = x^3 - 5$ is the function f shifted down 5 units, $g(x) = f(x) - 5$.

In addition to moving up and down, it is possible for functions to be vertically compressed and stretched. If $f(x) = \sqrt{x}$, then we have the following pairings

x	$f(x)$
0	0
4	2
9	3

Suppose we want to compress the vertical values by a factor of $\frac{1}{3}$. This would involve multiplying all of the vertical outputs of f by $\frac{1}{3}$. The adjustment to our table would be

x	$\frac{1}{3}f(x)$
0	0
4	$\frac{2}{3}$
9	1

where the function $\frac{1}{3}f(x) = \frac{1}{3}\sqrt{x}$, and we could call this new function g. The function g is the function f compressed by a factor of $\frac{1}{3}$, $g(x) = \frac{1}{3}f(x)$.

For $f(x) = \frac{1}{x}$, we can vertically stretch f by a factor of 4 if we multiply all of the vertical output values by 4.

x	$f(x)$
1	1
2	$\frac{1}{2}$
3	$\frac{1}{3}$

x	$4f(x)$
1	4
2	2
3	$\frac{4}{3}$

We can call this new function g where $g(x) = 4f(x) = \frac{4}{x}$.

The final type of vertical transformation is a reflection across the x-axis. This transformation involves changing the sign of all the vertical output values of a function. When all the vertical values experience a sign change, the function will reflect to the opposite side of the horizontal axis. This is why we refer to this transformation as a reflection over the x-axis. To see this transformation in action, let $f(x) = |x|$. Comparing the table of f to the table of the transformation of f would give us

x	$f(x)$
-3	3
0	0
3	3

x	$-f(x)$
-3	-3
0	0
3	-3

We could call this new function g and $g(x) = -|x| = -f(x)$.

3.5. VERTICAL TRANSFORMATIONS

The particular functions and transformations that have been presented serve only as examples. Vertical transformations can be applied to any function, and they modify the output of the function. If f is a function, then the vertical transformations on f are as follows:

$-f(x)$ Reflection across the x-axis

$cf(x)$ If $c > 1$, this is a vertical stretch by a factor of c

         If $0 < c < 1$, this is a vertical compression by a factor of c

$f(x) + c$ Vertical shift up c units

$f(x) - c$ Vertical shift down c units

A very useful skill involving transformations is to rewrite a given function as a transformation of another. As an example, suppose we were given the function $g(x) = \frac{2}{3x}$. We can rewrite $g(x)$ as

$$g(x) = \frac{2}{3x} = \left(\frac{2}{3}\right)\left(\frac{1}{x}\right) = \frac{2}{3}f(x)$$

where $f(x) = \frac{1}{x}$. The function g is the function $f(x) = \frac{1}{x}$ compressed vertically by a factor of $\frac{2}{3}$.

3.5.1. Multiple Vertical Transformations.
Often, there will be multiple vertical transformations on a given fundamental function. Consider the function

$$g(x) = \frac{3}{7}\sqrt{x} - 9$$

The function g is the result of two vertical transformations on the fundamental function $f(x) = \sqrt{x}$. To help us identify exactly what the vertical transformations on f might be, let us rewrite g as

$$g(x) = \frac{3}{7}f(x) - 9$$

This rewriting helps to verify that the two vertical transformations are a compression of $\frac{3}{7}$ and a shift down of 9. Furthermore, we can see from this equation that we need to perform the compression (multiply by $\frac{3}{7}$) before the shift (subtract 9).

It is critical that multiple vertical transformations be performed in the correct order. To help illustrate this fact, we know that $f(x) = \sqrt{x}$ contains the pairings

x	$f(x)$
0	0
1	1
4	2

and that the properly transformed pairings should satisfy the relationship

$$g(x) = \frac{3}{7}\sqrt{x} - 9 = \frac{3}{7}f(x) - 9$$

We begin by compressing the vertical values in the table by a factor of $\frac{3}{7}$.

x	$\frac{3}{7}f(x)$
0	0
1	$\frac{3}{7}$
4	$\frac{6}{7}$

and then shift these compressed vertical values down 9.

x	$\frac{3}{7}f(x) - 9$
0	-9
1	$-\frac{60}{7}$
4	$-\frac{57}{7}$

Note that each of these pairings satisfies the equation for $g(x) = \frac{3}{7}\sqrt{x} - 9$.

If we attempt to perform the vertical transformations in the other order (down 9 and then compress by a factor of $\frac{3}{7}$) the resulting pairings would be

x	
0	$-\frac{27}{7}$
1	$-\frac{24}{7}$
4	-3

none of which satisfies the equation for g.

In short, the correct order in which to transform the function $f(x) = \sqrt{x}$ into

$$g(x) = \frac{3}{7}\sqrt{x} - 9$$

3.5. VERTICAL TRANSFORMATIONS

is compress by $\frac{3}{7}$, then shift down 9. According to the order of operations, we must multiply (perform the compression) before we subtract (perform the shift down).

To further illustrate the correct order of multiple vertical transformations, suppose we rewrite $g(x) = \frac{3}{7}\sqrt{x} - 9$ as

$$g(x) = \frac{3}{7}\left(\sqrt{x} - 21\right)$$

Factoring out the $\frac{3}{7}$ does not change the function g, or any of it's pairings. This is still the same function we had before factoring. What does change is how we would transform $f(x) = \sqrt{x}$ to arrive at g. Substituting $f(x)$ in for $\sqrt{x}$ we would have

$$g(x) = \frac{3}{7}\left(f(x) - 21\right)$$

Remember, $f(x)$ is just a vertical output value of f. This means we would subtract 21 from any particular vertical value (which would correspond to a shift down of 21) and then multiply by $\frac{3}{7}$ (which would correspond to a vertical compression of $\frac{3}{7}$). Note that the order of these transformations (as well as the shift down) have changed, but the final transformed pairs do not.

Regardless of how a function is written, multiple vertical transformations are always performed by following how the order of operations dictates we must act on the output of the starting fundamental function.

Exercises for Section 3.5.

(1) Let the functions f, g, and h be defined as follows

$$f(x) = \sqrt{x} \qquad g(x) = \frac{1}{x} \qquad h(x) = x^3$$

Express each of the following as vertical transformations on f, g, or h. As an example, the function $r(x) = x^3 + 4$ would be $r(x) = h(x) + 4$.

(a) $j(x) = \sqrt{x} - 7$

(b) $k(x) = \dfrac{9}{x}$

(c) $l(x) = -x^3$

(d) $m(x) = 1 + \dfrac{2}{3}\sqrt{x}$

(e) $n(x) = \dfrac{1}{5}x^3 - 11$

(f) $p(x) = -\dfrac{7}{2x} - 8$

(2) For each function f, g, and h (as given in the previous problem), write out the equations that apply the following transformations in the order they are given.

   (a) A shift up of nine, then a compression by a factor of one third.

   (b) A compression by a factor of one third, then a shift up of nine.

   How far up will each of these graphs appear to be shifted? Why?

(3) The following functions are each some number of vertical transformations on one of the fundamental functions. List the starting fundamental function, and the transformations in the correct order as they are written. As an example, the function

$$g(x) = 2x^2 - 1$$

is the fundamental function $f(x) = x^2$ with two vertical transformations. The first is a stretch by a factor of two, the second is a shift down by one.

   (a) $j(x) = \dfrac{5}{7}\sqrt{x} - 7$

   (b) $k(x) = -\dfrac{1}{x} + 6$

   (c) $l(x) = \dfrac{11}{2}x^3 + 3$

   (d) $m(x) = 5|x| + \dfrac{4}{9}$

   (e) $n(x) = \dfrac{7}{3}\left(x^2 + 1\right)$

   (f) $p(x) = -\dfrac{9}{8}\left(\sqrt{x} - 1\right)$

   (g) $q(x) = -\dfrac{6}{17}x^2 + 1$

   (h) $r(x) = -\dfrac{14}{3}\left(\dfrac{1}{x} + 4\right)$

   Once you have completed all parts of this problem, use graphing software to help visualize your answers. The graphs and your answers should agree.

3.6. Horizontal Transformations

In a similar fashion to vertical transformations, the graph of any arbitrary function can be shifted left, right, horizontally compressed, horizontally stretched, or rotated about the vertical axis.

Let us begin with $f(x) = x^2$. We wish to shift this parabola 4 units to the right. This would mean all of our horizontal values need to be increased by 4, while the

3.6. HORIZONTAL TRANSFORMATIONS

vertical values remain the same. Following the same pattern from the previous section, the transformed parabola will be named g. The tables illustrating this horizontal shift is shown below.

x	$f(x)$
-2	4
0	0
2	4

x	$g(x)$
2	4
4	0
6	4

Given that the table for g is correct (the horizontal values are all shifted to the right by 4, while the vertical values remain unchanged), what is the equation for $g(x)$? We have seen that vertical transformations are the result of modification to the output of a function. Horizontal transformations are the result of modification to the input of a function. Rather than modifying the output of f with something like $g(x) = f(x) + 4$ (which is a vertical transformation), we need to modify the input of f. That is to say we need

$$g(x) = f(\text{some modification to } x)$$

We are trying to shift f to the right by 4 units. A natural thought would be $g(x) = f(x+4)$. This modification would leave us with

$$g(x) = f(x+4) = (x+4)^2$$

If we check this equation against the values for g from above we have conflicting results. The table tells us that $g(2) = 4$ while our potential equation for g tells us that $g(2) = (2+4)^2 = 6^2 = 36$. The values for $g(4)$ and $g(6)$ disagree as well (check for yourself). Our attempt at horizontally transforming f did not go as planned. Why?

The equation we guessed for g would correspond to the following pairings.

x	$g(x) = (x+4)^2$
-6	4
-4	0
-2	4

Compare this table to the one for f. This is a horizontal transformation of f, but it is a horizontal shift left by 4 units. Adding 4 to the input of f resulted in a horizontal shift to the left by 4. This left shift is a result of the fact that, for the same output value, the inputs to g are 4 less (or 4 to the left) than the inputs of f. As an example, the input of 6 to f, and the input of 2 (four less) to g result in the same output.

$$g(2) = (2+4)^2 = 36 \quad \text{while} \quad f(6) = (6)^2 = 36$$

To achieve our shift to the right, we need to subtract 4 from the input of f. This would require the input to be 4 larger (or 4 to the right of its previous value) to achieve the same vertical output. The correct equation for g is then

$$g(x) = f(x-4) = (x-4)^2$$

A quick check with the initial table for g verifies that this is our desired result.

For the next example, let $f(x) = x^3$. A horizontal transformation of 5 units to the left would be given by

$$g(x) = f(x+5) = (x+5)^3$$

Adding a positive number to the input of a function results in a shift to the left, while subtracting results in a shift to the right.

Horizontal stretching and compression are brought about by multiplication to a function's input. Let $f(x) = \sqrt{x}$, and let

$$g(x) = f(3x) = \sqrt{3x}$$

Some of the pairings of f and g would be

x	$f(x)$
1	1
4	2
9	3

x	$g(x)$
$\frac{1}{3}$	1
$\frac{4}{3}$	2
3	3

The vertical values in both tables are the same, while the horizontal values are not. The horizontal values of g have been compressed by a factor of $\frac{1}{3}$. The reason for this compression is that the input values of g are now being multiplied by 3 before the square root is taken. This multiplication by 3 means that the input values of g only need to be $\frac{1}{3}$ of the input values of f in order to achieve the same output value.

Keeping f as $f(x) = \sqrt{x}$ and letting

$$g(x) = f(\frac{1}{2}x) = \sqrt{\frac{1}{2}x}$$

would result in the following pairings.

x	$f(x)$
1	1
4	2
9	3

x	$g(x)$
2	1
8	2
18	3

For the same output values, the input values of g are double the input values f. The function g is the function f horizontally stretched by a factor of 2. This stretch is the due to the multiplication by $\frac{1}{2}$ before the square root is taken.

The final horizontal transformation we will discuss is a reflection across the vertical axis. This transformation involves changing the sign of all the horizontal input values of a function. When all of the horizontal values experience a sign change, the function will reflect to the opposite side of the vertical axis. To see this transformation in action we will keep $f(x) = \sqrt{x}$ and let

$$g(x) = f(-x) = \sqrt{-x}$$

Comparing some of the pairings of f and g would give us

x	$f(x)$
1	1
4	2
9	3

x	$g(x)$
-1	1
-4	2
-9	3

The function g is the function f reflected across the vertical axis.

The particular functions and transformations that have been covered serve only as examples. Horizontal transformations can be applied to any function, and they modify the input of the function. If f is a function, then the horizontal transformations on f are as follows:

$f(-x)$ Reflection across the vertical-axis

$f(cx)$ If $c > 1$, this is a horizontal compression by a factor of $\dfrac{1}{c}$

If $0 < c < 1$, this is a horizontal stretch by a factor of $\dfrac{1}{c}$

$f(x + c)$ Horizontal shift left c units

$f(x - c)$ Horizontal shift right c units

3.6.1. Multiple Horizontal Transformations.

As was the case with vertical transformations, being able to rewrite a given function as a horizontal transformation of another is a useful skill that warrants practice. Suppose that you are given a function

$$g(x) = \left(\frac{2}{3}x + 6\right)^3$$

Utilizing the fundamental function $f(x) = x^3$, we can rewrite g as

$$g(x) = \left(\frac{2}{3}x + 6\right)^3 = f\left(\frac{2}{3}x + 6\right)$$

This rewriting is helpful to illustrate the fact that the function g is the function f with two horizontal transformations applied. But, what exactly are they? It appears that the two transformations might be a horizontal stretch by a factor of $\frac{3}{2}$ and a shift left of 6. If that is correct (and we have not yet arrived at that conclusion), does the order in which we perform these two horizontal transformations make a difference?

As previously mentioned, horizontal transformations do not change a function's output. This means that we can select any output value we might want, and solve for the input associated with it. Consider the function $f(x) = x^3$ and the pairing that f makes with an input of 2.

$$f(2) = 2^3 = 8$$

Thus, we know that the point $(2, 8)$ is on the graph of f. As we know that g is some number of horizontal transformations on the function f, we can now consider what input value for g will result in an output of 8. All we need to do is set the equation for $g(x)$ equal to 8 and solve for the input.

$$8 = \left(\frac{2}{3}x + 6\right)^3$$

$$2 = \frac{2}{3}x + 6$$

$$-4 = \frac{2}{3}x$$

$$-6 = x$$

The function g pairs an input of -6 with an output of 8, which is to say that the point $(-6, 8)$ is on the graph of g. Consider how we solved for x. After taking the cubed root (which was not part of any transformation), we subtracted 6 and then multiplied by $\frac{3}{2}$. The subtraction of 6 was the left shift of 6, and the multiplication by $\frac{3}{2}$ was the horizontal stretch by a factor of $\frac{3}{2}$. The order in which we performed those operations (shift then stretch) was the correct order in which to perform the transformations.

Given our two functions

$$f(x) = x^3 \quad \text{and} \quad g(x) = \left(\frac{2}{3}x + 6\right)^3$$

it would not be correct to say that the two horizontal transformations going from f to g are a stretch by $\frac{3}{2}$ and a shift left of 6. They are instead a shift left of 6, and a stretch by $\frac{3}{2}$. The order of the transformations is critical.

To further illustrate this idea, suppose we rewrite the function g as

$$g(x) = \left(\frac{2}{3}x + 6\right)^3 = \left(\frac{2}{3}(x + 9)\right)^3$$

where the only difference is that the $\frac{2}{3}$ has been factored out. Once again, we will set the output $g(x)$ equal to 8 and solve for the input.

$$8 = \left(\frac{2}{3}(x + 9)\right)^3$$

$$2 = \frac{2}{3}(x + 9)$$

$$3 = x + 9$$

$$-6 = x$$

That we arrive at $x = -6$ is not interesting (g is still the same function). What is interesting are the algebraic steps, and their horizontal transformation interpretations. After taking the cubed root (which was not a part of any transformation), we multiplied by $\frac{3}{2}$ and then subtracted 9. The multiplication by $\frac{3}{2}$ was a horizontal stretch by $\frac{3}{2}$, and the subtraction of 9 was a left shift of 9. Rewriting the equation for $g(x)$ changed the algebra required to solve for x, and in turn changed the horizontal transformations.

Given our two functions

$$f(x) = x^3 \quad \text{and} \quad g(x) = \left(\frac{2}{3}(x + 9)\right)^3$$

the two horizontal transformations going from f to g are a stretch by $\frac{3}{2}$ and a shift left of 9, and they must be performed in that order.

To summarize, there are two equivalent and correct ways to consider the transformations between the functions

$$f(x) = x^3 \quad \text{and} \quad g(x) = \left(\frac{2}{3}x + 6\right)^3 = \left(\frac{2}{3}(x+9)\right)^3$$

The first is to say that g is two horizontal transformations on f; a shift left of 6, followed by a horizontal stretch of $\frac{3}{2}$. The second is to say that g is two horizontal transformations on f; a horizontal stretch by $\frac{3}{2}$, followed by a shift left of 9. Both sets of transformations arrive at all of the same pairings, and in both cases the order in which the transformations must be performed is dictated by how you solve for x.

The graph of g is shown below. Note that the overall shift appears to be 9 to the left. This is a direct result of the order in which the horizontal transformations are applied. If you are shifting before stretching, the shift will have the stretch applied to it (i.e $-6 \cdot \frac{3}{2} = -9$). If you are stretching before shifting, the shift will appear exactly as written in the function.

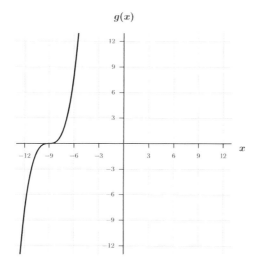

FIGURE 12. The function $g(x) = \left(\frac{2}{3}x + 6\right)^3 = \left(\frac{2}{3}(x+9)\right)^3$.

Exercises for Section 3.6.

(1) Let the functions f, g, and h be defined as follows

$$f(x) = \sqrt{x} \qquad g(x) = \frac{1}{x} \qquad h(x) = x^2$$

Express each of the following as horizontal transformations on f, g, or h. As an example, the function $r(x) = (x+4)^2$ would be $r(x) = h(x+4)$.

(a) $j(x) = \sqrt{x+12}$

(b) $k(x) = \dfrac{1}{x-9}$

(c) $l(x) = (-x)^2$

(d) $m(x) = \sqrt{\dfrac{2}{5}x+3}$

(e) $n(x) = \left(\dfrac{6}{13}x - 11\right)^2$

(f) $p(x) = \dfrac{1}{2x-7}$

(2) For each function f, g, and h (as given in the previous problem), write out the equations that apply the following horizontal transformations in the order they are given.

(a) A shift right of nine, then a compression by a factor of one third.

(b) A compression by a factor of one third, then a shift right of nine.

How far to the right will each of these graphs appear to be shifted? Why?

(3) The following functions are each some number of horizontal transformations on one of the fundamental functions. List the starting fundamental function, and the transformations in the correct order as they are written. As an example, the function

$$g(x) = (2x+3)^2$$

is the fundamental function $f(x) = x^2$ with two horizontal transformations. The first is a left shift by three, the second is a compression by a factor of one half.

(a) $j(x) = \sqrt{4x+1}$

(b) $k(x) = \dfrac{1}{-x+6}$

(c) $l(x) = \left(\dfrac{1}{2}x - 9\right)^3$

(d) $m(x) = |5x - 2|$

(e) $n(x) = \left(\dfrac{3}{2}x + 4\right)^2$

(f) $p(x) = \sqrt{\dfrac{4}{7}(x-3)}$

(g) $q(x) = \left[-\left(x - \dfrac{7}{9}\right)\right]^2$ (h) $r(x) = \dfrac{1}{-\left(x + \frac{6}{17}\right)}$

Once you have completed all parts of this problem, use graphing software to help visualize your answers. The graphs and your answers should agree.

3.7. Vertical and Horizontal Transformations

Vertical and horizontal transformations do not interact. This means that you are free to perform them in any order, relative to one another, you choose. However, you must still perform all of the vertical transformations in their respective correct order, and all of the horizontal transformations in their respective correct order.

As an example, consider the function

$$g(x) = \dfrac{4}{7}\sqrt{2x - 5} + 1$$

This function is a transformation on the fundamental square root function, $f(x) = \sqrt{x}$. Specifically, g is f with two vertical and two horizontal transformations applied. In their correct respective orders, those transformations are

Horizontal

(1) Shift right 5
(2) Compress by a factor of $\frac{1}{2}$

Vertical

(1) Compress by a factor of $\frac{4}{7}$
(2) Shift up 1

We can transform pairings of f into pairings of g as long as we perform the horizontal and vertical transformations in their (respectively) correct order. Starting with three randomly selected pairs of f, the horizontal transformations would be

x	$f(x)$
1	1
4	2
9	3

$\rightarrow$

x	$f(x - 5)$
6	1
9	2
14	3

$\rightarrow$

x	$f(2x - 5)$
3	1
$\frac{9}{2}$	2
7	3

Note that none of the vertical output values have changed. Starting from the completed horizontal, the vertical would give us

3.7. VERTICAL AND HORIZONTAL TRANSFORMATIONS

x	$f(2x-5)$
3	1
$\frac{9}{2}$	2
7	3

→

x	$\frac{4}{7}f(2x-5)$
3	$\frac{4}{7}$
$\frac{9}{2}$	$\frac{8}{7}$
7	$\frac{12}{7}$

→

x	$g(x) = \frac{4}{7}f(2x-5)+1$
3	$\frac{11}{7}$
$\frac{9}{2}$	$\frac{15}{7}$
7	$\frac{19}{7}$

When trying to determine if a transformation is horizontal or vertical, remember the following: horizontal transformations involve modification to the input of a function, while vertical transformations involve modification to the output of a function.

Exercises for Section 3.7.

(1) Express each of the following as some number of horizontal and/or vertical transformations on one of the fundamental functions. Be sure to put the transformations in their respective correct orders.

(a) $j(x) = \frac{2}{3}\sqrt{x-6}$

(b) $k(x) = \dfrac{5}{x+7}$

(c) $l(x) = \frac{1}{7}(-x+2)^2 - 8$

(d) $m(x) = \frac{3}{2}\sqrt{\frac{3}{2}x - 4} + 1$

(e) $n(x) = \frac{4}{13}(5x-1)^2 - 4$

(f) $p(x) = \dfrac{-1}{7(x-11)} + 9$

(g) $q(x) = -6(-x+1)^3 + 3$

(h) $r(x) = 12\left|\frac{1}{6}(x+2)\right| + \frac{11}{5}$

(2) Suppose you have the graph of a function f. Determine the transformations on f such that the function

$$g = \text{some number of transformations on } f$$

is symmetric to f across $x = 3$.

(3) The graph of the function m is shown below. You may assume that plotted points have integer value coordinates. Use the graph of m as a starting point to sketch each of the following.

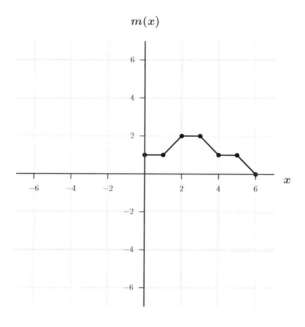

(a) $2m(-x)$ (c) $-\frac{1}{2}m(x)$ (e) $m(\frac{1}{3}x + 2)$

(b) $m(x+1) - 2$ (d) $m(2x) + 1$ (f) $-m(-x) - 1$

(4) Each of the graphs below are some number of transformations on one of the fundamental functions. Determine the equation of each function. You may assume plotted points have integer value coordinates.

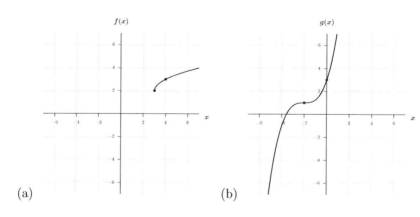

(a)   (b)

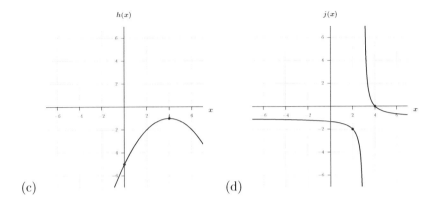

(c) (d)

3.8. Composition

This section will expand upon our previous discussion of function composition. Specifically, we will explore the idea of using one function as input to another, and provide some concrete examples of the process.

Let f and g be functions such that

$$f(x) = x^2 - 3 \quad \text{and} \quad g(x) = 6x + 1$$

The notation $f(g(x))$ represents the composition f of g (if this doesn't ring any bells, you should revisit the earlier section on the theory of function composition). Evaluating $f(g(x))$ is analogous to evaluating something like $f(4)$. To evaluate $f(4)$ we would replace every x in the expression $x^2 - 3$ with a 4 and then simplify. To evaluate $f(g(x))$, replace every x in $x^2 - 3$ with $6x + 1$ (note that this is the equation for the pairings of g) and simplify.

$$f(g(x)) = (6x + 1)^2 - 3$$
$$= 36x^2 + 12x + 1 - 3$$
$$= 36x^2 + 12x - 2$$

To evaluate g of f, or $g(f(x))$, we would replace every x in the expression $6x + 1$ with $x^2 - 3$ (the equation for the pairings of f) and then simplify.

$$g(f(x)) = 6(x^2 - 3) + 1$$
$$= 12x^2 - 18 + 1$$
$$= 12x^2 - 17$$

The domains of $f(g(x))$ and $g(f(x))$ are both $\mathbb{R}$. Note that the compositions are not equal. That is to say $f(g(x)) \neq g(f(x))$. Function composition is not necessarily commutative.

As another example, let f and g be given by

$$f(x) = \frac{2}{3x - 5} \quad \text{and} \quad g(x) = \sqrt{x - 9}$$

The composition $f(g(x))$ is then

$$f(g(x)) = \frac{2}{3\sqrt{x - 9} - 5}$$

What is the domain of this composition? We know that the radicand must be at least zero, and that the denominator must not be zero. If $x \geq 9$, then the radicand is at least zero. To determine where the denominator is equal to zero, set it equal to zero and solve.

$$0 = 3\sqrt{x - 9} - 5$$

$$5 = 3\sqrt{x - 9}$$

$$\frac{5}{3} = \sqrt{x - 9}$$

$$\frac{25}{9} = x - 9$$

$$\frac{25}{9} + 9 = x$$

$$\frac{106}{9} = x$$

The domain for the composition $f(g(x))$ is then $\{x \mid x \in \mathbb{R} \text{ and } x \geq 9 \text{ and } x \neq \frac{106}{9}\}$.

Evaluating $g(f(x))$ would give us the following

$$g(f(x)) = \sqrt{\frac{2}{3x - 5} - 9}$$

Always identify the domain of a composition before simplification. Remember, a function is defined by exactly what is given, not by how it may be simplified. We know that the denominator of the fraction cannot be equal to zero, so $x \neq \frac{5}{3}$. Additionally, we need the radicand to be at least zero. A sign chart will allow us to determine where this occurs.

3.8. COMPOSITION

$$0 = \sqrt{\frac{2}{3x-5} - 9}$$

$$0 = \frac{2}{3x-5} - 9$$

$$9 = \frac{2}{3x-5}$$

$$9(3x-5) = 2$$

$$3x - 5 = \frac{2}{9}$$

$$3x = \frac{2}{9} + 5$$

$$3x = \frac{47}{9}$$

$$x = \frac{47}{27}$$

The sign chart based on the undefined point and the zero is then given by

		$\frac{5}{3}$		$\frac{47}{27}$	
$\frac{2}{3x-5} - 9$	$-$	UN	$+$	0	$-$

From the sign chart we know that the domain of $g(f(x))$ is then $\{x \mid x \in \mathbb{R} \text{ and } \frac{5}{3} < x \leq \frac{47}{27}\}$. Simplification of $g(f(x))$ can now be carried out.

$$g(f(x)) = \sqrt{\frac{2}{3x-5} - 9}$$

$$= \sqrt{\frac{2}{3x-5} - \frac{9(3x-5)}{3x-5}}$$

$$= \sqrt{\frac{2 - 27x + 45}{3x-5}}$$

$$= \sqrt{\frac{47 - 27x}{3x-5}}$$

As a final example, let

$$f(x) = \frac{1}{x} \quad \text{and} \quad g(x) = \frac{1}{2x-3} + 1$$

The composition $f(g(x))$ would be

$$f(g(x)) = \frac{1}{\frac{1}{2x-3} + 1}$$

Again, find the domain before simplification. No denominators can equal zero, so

$$2x - 3 \neq 0 \quad \text{and} \quad \frac{1}{2x-3} + 1 \neq 0$$

Solving the first of these inequalities tells us that $x \neq \frac{3}{2}$, and the second gives $x \neq 1$. The domain of the composition $f(g(x))$ is then $\{x \mid x \in \mathbb{R} \text{ and } x \neq \frac{3}{2} \text{ and } x \neq 1\}$. Simplification would be done as follows.

$$f(g(x)) = \frac{1}{\frac{1}{2x-3} + 1}$$

$$= \frac{1}{\frac{1}{2x-3} + \frac{2x-3}{2x-3}}$$

$$= \frac{1}{\frac{2x-2}{2x-3}}$$

$$= \frac{2x-3}{2x-2}$$

Note that the final simplified version of $f(g(x))$ does not show that x can not equal $\frac{3}{2}$, it only shows that $x \neq 1$. The simplification process (while algebraically correct) removed our ability to determine one of the function's undefined points. This is exactly why you must determine the domain of a composition before simplification.

Writing out $g(f(x))$ would give us

$$g(f(x)) = \frac{1}{2\left(\frac{1}{x}\right) - 3} + 1$$

The domain of this composition is all real numbers that do not result in a denominator being equal to zero. This means that

$$x \neq 0 \quad \text{and} \quad 2\left(\frac{1}{x}\right) - 3 \neq 0$$

Solving the second for x gives $x \neq \frac{2}{3}$. The domain of $g(f(x))$ is then $\{x \mid x \in \mathbb{R} \text{ and } x \neq 0 \text{ and } x \neq \frac{2}{3}\}$.

Simplification of $g(f(x))$ is given by

$$g(f(x)) = \frac{1}{2\left(\frac{1}{x}\right) - 3} + 1$$

$$= \frac{1}{\frac{2}{x} - \frac{3x}{x}} + 1$$

$$= \frac{1}{\frac{2-3x}{x}} + 1$$

$$= \frac{x}{2 - 3x} + 1$$

$$= \frac{x}{2 - 3x} + \frac{2 - 3x}{2 - 3x}$$

$$= \frac{2 - 2x}{2 - 3x}$$

As a final reminder, note that the simplified version of $g(f(x))$ fails to show that $x \neq 0$. Always determine the domain of a composition before you simplify.

Exercises for Section 3.8.

(1) Let the functions f, g, and h be defined as follows

$$f(x) = \sqrt{x} \qquad g(x) = \frac{1}{x} \qquad h(x) = x^2$$

Algebraically evaluate each of the following.

(a) $g(x - 1)$

(b) $f\left(\dfrac{x+9}{x+2}\right)$

(c) $g\left(\dfrac{2}{5x}\right)$

(d) $h(-x + 2)$

(e) $f(x^2 + 10x + 25)$

(f) $h\left(\dfrac{x+6}{\sqrt{x+6}}\right)$

(2) The functions f and g are defined by

x	0	a	2	b
$f(x)$	b	a	0	2
$g(x)$	a	2	b	0

Evaluate each of the following

(a) $f(g(2))$

(c) $f(g(b))$

(e) $f(f(2))$

(b) $g(f(a))$

(d) $g(f(0))$

(f) $g(g(b))$

(3) Evaluate $f(g(x))$ and $g(f(x))$ for each of the following. Be sure to state the domain of each composition.

(a) $f(x) = \dfrac{1}{x}$ $g(x) = \sqrt{x-4}$

(b) $f(x) = 4x^2$ $g(x) = \sqrt{2x-1}$

(c) $f(x) = \dfrac{2}{x+3}$ $g(x) = \dfrac{5}{x^2}$

(d) $f(x) = \sqrt{9-x}$ $g(x) = \dfrac{1}{x^2}$

(4) The functions f, g, and h are defined as follows:

$$f(x) = \dfrac{1}{x} \qquad g(x) = x^2 \qquad h(x) = \sqrt{x} + 2$$

Express each of the following as a composition involving f, g, or h.

(a) $j(x) = x^4$

(c) $l(x) = |x| + 2$

(e) $n(x) = x + 4\sqrt{x} + 4$

(b) $k(x) = \sqrt{\dfrac{1}{x} + 2}$

(d) $m(x) = x$

(f) $p(x) = \sqrt{\sqrt{x} + 2} + 2$

(5) Suppose we have the relationship that

$$h(x) = f(g(x))$$

and that all pairs of the functions f and h are given by

x	$f(x)$
8	7
9	a
10	3
11	9
12	2

x	$h(x)$
0	a
1	3
4	7
5	2
6	9

Write out all pairings of the function g

3.9. Composition and Graphing

The ideas of function composition can be helpful in the process of graphing many different types of functions. As an example, consider the function

$$f(x) = \sqrt{x^2 - 5x - 6}$$

A direct approach to graphing this function would be to choose input values for x, plug them into the equation for $f(x)$, and then plot the resulting points, $(x, f(x))$, on a coordinate plane. This approach is not incorrect, and would ultimately yield the correct graph of f. However, given what we know about function composition, there is another way to graph f.

It is often possible to take a single, somewhat complicated function and *decompose* it into several simpler functions. The function f could be thought of as a composition of the functions g and h where

$$g(x) = \sqrt{x} \quad \text{and} \quad h(x) = x^2 - 5x - 6$$

which would give us

$$f(x) = \sqrt{x^2 - 5x - 6} = g\left(h(x)\right)$$

Take a moment and consider what this relationship is telling us. In short, we would have the following order of events.

(1) Choose an input, we will call it x.

(2) Calculate the output from h, based on an input of x. In other words, $h(x)$.

(3) This output, $h(x)$, then becomes an input into the function g.

(4) The output of g, based on an input of $h(x)$, $g(h(x))$, is equal to the output of f based on an input of x. That is to say, $f(x) = g(h(x))$.

These steps become more apparent if we use some actual numbers to illustrate the process. Suppose we let $x = -3$. On the left we will calculate $h(-3)$ and then $g(h(-3))$, while on the right we will calculate $f(-3)$.

$$h(3) = (-3)^2 - 5(-3) - 6$$
$$= 9 + 15 - 6$$
$$= 18$$
$$g(18) = \sqrt{18}$$
$$= 3\sqrt{2}$$

$$f(-3) = \sqrt{(-3)^2 - 5(-3) - 6}$$
$$= \sqrt{9 + 15 - 6}$$
$$= \sqrt{18}$$
$$= 3\sqrt{2}$$

Just as expected, we have $f(-3) = g(h(-3)) = 3\sqrt{2}$.

The fact that we are using outputs of the function h as inputs to the function g is the basis of this entire section. Consider a graph of the function h.

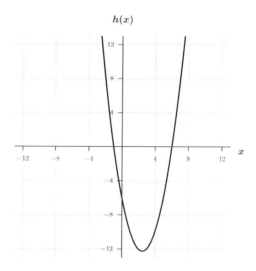

FIGURE 13. The function $h(x) = x^2 - 5x - 6$.

The vertical values of this graph represent outputs of the function h. We are going to use these vertical values as input to the function $g(x) = \sqrt{x}$. Based on our knowledge of the square root function, we immediately know the graph of

$$f(x) = g(h(x))$$

cannot exist between the x-intercepts of the graph shown in Figure 13. All of the vertical values (output values of h) are negative between the x-intercepts, and we cannot use a negative number as input to the square root function.

Furthermore, all of the output values on either side of the x-intercepts are positive increasing numbers, and we are already familiar with graphing the square root of positively increasing numbers. Based only on our previous knowledge of some fundamental functions, analyzing this problem through function composition has provided us with a good mental sketch as to the graph of f. The graph of $f(x) = g(h(x))$ is shown in Figure 14.

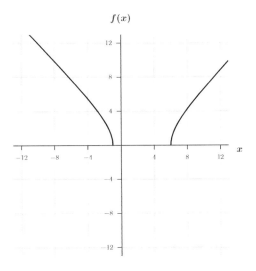

FIGURE 14. The function $f(x) = g(h(x)) = \sqrt{x^2 - 5x - 6}$.

Exercises for Section 3.9.

(1) Decompose each of the following functions into a composition involving two simpler functions. Sketch a graph of the inner function and state its domain, then sketch a graph of the final composition and state the domain of the final composition.

(a) $f(x) = \sqrt{x^2 - 4}$

(b) $g(x) = \dfrac{1}{x^2}$

(c) $h(x) = |(x-1)^3|$

(d) $j(x) = (|x| - 1)^3$

(e) $k(x) = (\sqrt{x})^2 - 2$

(f) $l(x) = \dfrac{1}{\sqrt{x}}$

(2) Compare functions h and j from the previous problem, and note how the placement of the absolute value bars resulted in significantly different results. One possible decomposition for the function j would be $j(x) = f(g(x))$ where $f(x) = (x-1)^3$ and $g(x) = |x|$. These choices would result in

$$j(x) = f(|x|)$$

Stepping outside this particular example for a moment, whenever you are able to write a composition where the inner function is the absolute value function, the final composition is even. The proof of this idea is fairly straightforward, regardless of the functions f and j, if $j(x) = f(|x|)$

$$j(x) = f(|x|) = f(x)$$
$$j(-x) = f(|-x|) = f(x)$$

and we have that $j(x) = j(-x)$.

Your graph for the function j in the previous problem should have even symmetry. Specifically, it should be the graph of the fundamental cubic for $x \geq 0$ shifted one to the right, and then reflected across the vertical axis.

Using algebra, prove that $j(x) = (|x| - 1)^3$ is even.

(3) Decompose each of the following such that the inner function is the absolute value function, then sketch the compositions.

(a) $f(x) = \sqrt{|x| + 3}$

(b) $g(x) = \dfrac{1}{|x| - 4}$

(c) $h(x) = ||x| - 3|$

(d) $j(x) = (|x| - 2)^2 - 5$

(4) The graph of function f is given below. Assume that the plotted point has coordinates $(0, 1)$. Use the graph of f to sketch each of the following. Be sure to label the transformation of the provided point.

(a) $g(x) = f(-x)$

(b) $h(x) = \dfrac{1}{f(x)}$

(c) $j(x) = |f(x)|$

(d) $k(x) = f(|x|)$

(e) $l(x) = \sqrt{f(x)}$

(f) $m(x) = (f(x))^2$

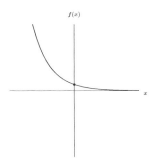

3.10. Inverses

Our previous work in function theory tells us that all bijective functions have inverses. Suppose we have a function f such that

$$f(x) = \frac{3}{7}x - 11$$

This function is bijective, so it must have an inverse. How might we determine the equation that shows us how to calculate the pairings of f^{-1} (the inverse of f)? It is perfectly reasonable to think that a function's inverse "undoes" the pairing process of the function. When working with functions whose pairing is determined via a mathematical equation, the idea of "undoing" the mathematical equation responsible for the pairings would mean reversing all of the algebraic operations that transformed the input value into the output value. In other words, solving for x.

The steps below outline a traditional approach to determining the equation of a function's inverse.

(1) Use a temporary dummy variable to replace $f(x)$.

(2) Solve for the dependent variable (the input of the original function).

(3) Replace the dummy variable with your new input variable and replace the independent variable you solved for with $f^{-1}(x)$.

For our function f defined above, we would do the following:

$$f(x) = \frac{3}{7}x - 11$$

$$y = \frac{3}{7}x - 11 \qquad \text{Use a dummy variable of } y \text{ to replace } f(x)$$

$$y + 11 = \frac{3}{7}x \qquad \text{Solve for } x$$

$$\frac{7}{3}(y+11) = x$$

$$\frac{7}{3}(x+11) = f^{-1}(x) \qquad \text{Replace } y \text{ with } x \text{ and } x \text{ with } f^{-1}(x)$$

Just like that, we have the equation for the pairings of f^{-1}, and we can consider f^{-1} as a function in its own right. In this example, the domain of f and f^{-1} are both $\mathbb{R}$. This is not always the case.

Consider the function

$$f(x) = 2x^2 - 8x + 3$$

This function presents somewhat of a problem. It is not bijective. Specifically, this parabola is not 1-1. In fact, no parabola are 1-1. With the exception of the vertex, all values in the range of a parabola are paired with two domain values (thus violating the definition of 1-1). As an example, suppose we wish to know what domain values this parabola pairs with the range value of 10.

$$10 = 2x^2 - 8x + 3$$
$$0 = 2x^2 - 8x - 7$$

From which the quadratic formula tells us

$$x = \frac{4 + \sqrt{30}}{2} \qquad \text{and} \qquad x = \frac{4 - \sqrt{30}}{2}$$

Recall that the domain of a function is the range of its inverse (and vice versa). Left unchecked, the inverse of f would treat these solutions as range values, and pair the domain value of 10 with both. Such a pairing would violate the definition of a function.

To avoid this problem, we will restrict the domain of f in such a fashion as to make it 1-1. It is important that we keep the range the same when we make this restriction on the domain. As presented, the domain of f is $\mathbb{R}$, and the range of f is $\{f(x) \mid f(x) \in \mathbb{R} \text{ and } f(x) \geq -5\}$. To make f 1-1, we will restrict the domain by cutting this parabola in half, vertically, through the vertex.

It makes no difference which half of the parabola we choose. The restricted domain of f can either be the left half of the parabola, $\{x \mid x \in \mathbb{R} \text{ and } x \leq 2\}$, or the right half of the parabola, $\{x \mid x \in \mathbb{R} \text{ and } x \geq 2\}$. Both options leave a range of $\mathbb{R}$, and a function that is 1-1. We will choose the right half with a domain of $\{x \mid x \in \mathbb{R} \text{ and } x \geq 2\}$.

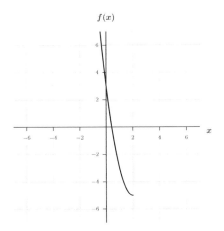

FIGURE 15. Domain
$\{x \mid x \in \mathbb{R} \text{ and } x \leq 2\}$

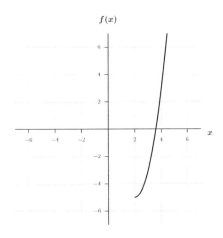

FIGURE 16. Domain
$\{x \mid x \in \mathbb{R} \text{ and } x \geq 2\}$

At this point, the calculation of the inverse is exactly as previously described.

$$f(x) = 2x^2 - 8x + 3$$

$$y = 2x^2 - 8x + 3$$

$$y = 2(x^2 - 4x + 4) + 3 - 8 \qquad \text{Complete the square}$$

$$y = 2(x - 2)^2 - 5$$

$$\frac{y + 5}{2} = (x - 2)^2$$

$$\pm\sqrt{\frac{y + 5}{2}} = x - 2$$

$$\sqrt{\frac{y + 5}{2}} = x - 2 \qquad \text{Choose the positive root}$$

$$\sqrt{\frac{y + 5}{2}} + 2 = x$$

$$\sqrt{\frac{x+5}{2}} + 2 = f^{-1}(x)$$

Pay close attention to the selection of the positive square root. Our domain choice tells us x must be at least 2, which means $x - 2$ has to be positive (thus requiring us to select the positive root). The domain of f^{-1} is $\{x \mid x \in \mathbb{R} \text{ and } x \geq -5\}$, while the range is $\{f^{-1}(x) \mid f^{-1}(x) \in \mathbb{R} \text{ and } f^{-1}(x) \geq 2\}$.

Graphing f and f^{-1} together on the same coordinate plane (Figure 17) reveals an important relationship. A function and its inverse are symmetric about the identity

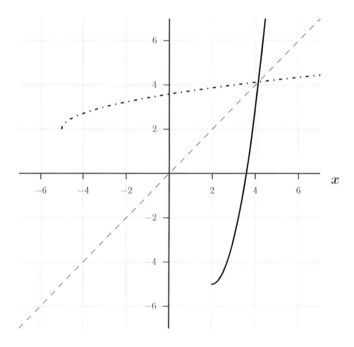

FIGURE 17. f is the solid line, f^{-1} is the dashed and dotted line, and $i(x) = x$ is the dashed line

function $i(x) = x$. This identity symmetry is due to the fact that the domain of a function is the range of its inverse, and vice versa. We can examine this relationship by considering a few of the pairings from the function, and the inverse we have been working with. For

$$f(x) = 2x^2 - 8x + 3 \quad \text{and} \quad f^{-1}(x) = \sqrt{\frac{x+5}{2}} + 2$$

a few randomly chosen pairings could be

x	$f(x)$
2	−5
3	−3
4	3

x	$f^{-1}(x)$
−5	2
−3	3
3	4

This swap between horizontal and vertical coordinates is responsible for the symmetry about the identity function.

A final thought worth considering is the composition of a function and its inverse. Our previous work in function theory tells us

$$f\left(f^{-1}(x)\right) = x \qquad \text{and} \qquad f^{-1}\left(f(x)\right) = x$$

We can verify either of these composition relationships with any real function via algebraic simplification. Using the parabola from our earlier example

$$f\left(f^{-1}(x)\right) = 2\left(\sqrt{\frac{x+5}{2}} + 2\right)^2 - 8\left(\sqrt{\frac{x+5}{2}} + 2\right) + 3$$

$$= 2\left(\frac{x+5}{2} + 4\sqrt{\frac{x+5}{2}} + 4\right) - 8\left(\sqrt{\frac{x+5}{2}} - 2\right) + 3$$

$$= x + 5 + 8\left(\sqrt{\frac{x+5}{2}}\right) + 8 - 8\left(\sqrt{\frac{x+5}{2}}\right) - 16 + 3$$

$$= x$$

where this composition is defined on the domain $\{x \mid x \in \mathbb{R} \text{ and } x \geq -5\}$. It would be good practice to verify the other composition, $f^{-1}(f(x))$, as an exercise.

As a final point of clarification, the composition $f\left(f^{-1}(x)\right) = x$ on the invertible domain of f^{-1}, not the domain of the composition. Additionally, the composition $f^{-1}(f(x)) = x$ on the invertible domain of f, not the domain of the composition. This fact is best illustrated with an example. Consider the functions

$$f(x) = x^2 \qquad \text{and} \qquad f^{-1}(x) = \sqrt{x}$$

The composition $f^{-1}(f(x))$ would be given by

$$f^{-1}(f(x)) = \sqrt{x^2}$$

The domain of this composition is $\mathbb{R}$. However, considering the square root function as the inverse of f requires a restriction on the domain of f (parabola are not 1-1). If we choose to restrict the domain of f to $\{x \mid x \in \mathbb{R} \text{ and } x \geq 0\}$, then we have the relationship $f^{-1}(f(x)) = x$ on a domain of $\{x \mid x \in \mathbb{R} \text{ and } x \geq 0\}$. The graph of the composition is given in Figure 18. This graph should look familiar. When we previously encountered this graph, we were considering the graph of the absolute value function. Note that

$$\sqrt{x^2} = \begin{cases} x & \text{if } x \geq 0 \\ -x & \text{if } x < 0 \end{cases}$$

which is to say that $\sqrt{x^2} = |x|$.

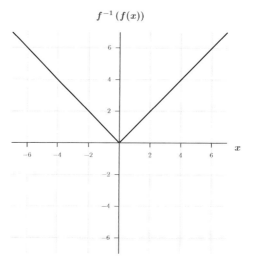

FIGURE 18. The composition $f^{-1}(f(x)) = \sqrt{x^2}$.

Exercises for Section 3.10.

(1) Determine the equation of the inverse function for each of the following. State the invertible domain and range of the original function, and the domain and range of the inverse.

(a) $f(x) = (x-5)^2 - 6$

(b) $g(x) = x^2 + 4x + 1$

(c) $h(x) = \dfrac{2}{x+3} - 5$

(d) $j(x) = (x-5)^3 + 7$

(e) $k(x) = -\sqrt{3-x} + 2$

(f) $l(x) = \sqrt{x^2 - 9}$

(2) Choose three parts of the previous problem to sketch. Be sure each sketch contains the function and its inverse.

(3) Suppose that the function T pairs an input of "number of beavers per acre" with an output, $T(b)$, that corresponds to the "number of trees chewed per acre". Explain, with units, each of the following.

(a) $T(2)$

(b) $T(12) = 35$

(c) $T^{-1}(54)$

(d) $T^{-1}(29) = 10$

(4) Given the following functions

$$f(x) = 2x^2 - 8x + 3 \quad \text{and} \quad f^{-1}(x) = \sqrt{\dfrac{x+5}{2}} + 2$$

Evaluate $f^{-1}(f(x))$.

(5) Explain why the graph of a function and its inverse will always exhibit identity symmetry.

(6) Suppose that the function f is given by

$$f(x) = ax^2 + bx + 7$$

and you know $f^{-1}(5) = 1$, and $f(3) = 2$. Solve for a and b.

(7) The graph of the function f is shown below. You may assume that plotted points have integer value coordinates. Use the graph of f to sketch the functions g and h as defined below.

(a) $g(x) = f^{-1}(x)$

(b) $h(x) = (f(x))^{-1}$

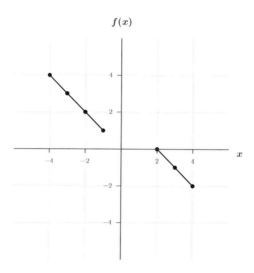

(8) The domain and range of the function f are both given by

$$\{-3, -2, -1, 0, 1, 2, 3\}$$

Complete the following table.

x	$f(x)$	$(f(x))^{-1}$	$f(-x)$	$f^{-1}(x)$
-3				
-2	-3		0	
-1				3
0		1		
1		$-\frac{1}{2}$		
2				
3			2	

CHAPTER 4

An Introduction to Mathematical Modeling

Mathematical modeling is the process of using mathematical equations to represent some real-life phenomena. A very well-known example of this process is weather prediction. How is it that meteorologists make a weather forecast? The short answer is that they take current atmospheric observations (temperature, humidity, barometric pressure, etc.) and plug those values into some extremely complicated mathematical equations. These equations represent our best efforts to recreate the fantastic complexity of the atmosphere with mathematics. In essence, these equations are a mathematical representation of how we think our atmosphere operates. A computer is then used to solve these equations into the future and make predictions about what we think might happen with upcoming weather patterns.

The mathematical models that represent our atmosphere are (more often than not) exceedingly complex. While we won't get anywhere near that level of complexity, we will consider several different types of mathematical models, starting with linear.

4.1. The Least Squares Regression Line

Consider the following scenario. For the seven exams in your precalculus class, you record the number of hours spent studying, and the score on the exam. Your collected data is given below.

Hours Studying	Exam Score
3	81
1	65
2	62
5	93
4	87
3.5	74
0.5	51

We can consider exam score as dependent on the amount of time spent studying. Thus we can treat time spent studying as the independent variable, score as the dependent variable, and graph this data.

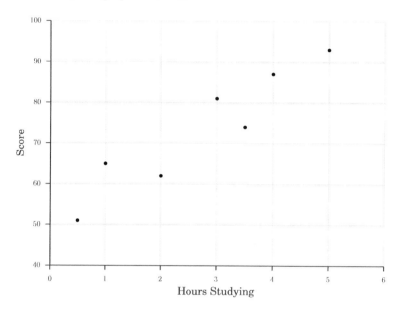

FIGURE 1. Collected testing data

A visual inspection of our data suggests a linear trend. That is to say that the data seems to resemble a straight line. The goal in developing a linear model for this data is to find the linear equation that best represents all the data points simultaneously. How might we approach finding such a line? A reasonable first attempt would be to find the line that connects the two most extreme data points $(0.5, 51)$ and $(5, 93)$. We begin by finding the slope between those two points.

$$m = \frac{93 - 51}{5 - 0.5}$$

$$= \frac{42}{4.5}$$

$$= \frac{28}{3}$$

Using the slope with either of the points, we can determine the equation of the line connecting them. We will use the point $(0.5, 51)$.

$$y - 51 = \frac{28}{3}(x - 0.5)$$

$$y - 51 = \frac{28}{3}x - \frac{14}{3}$$

$$y = \frac{28}{3}x + \frac{139}{3}$$

This equation represents our first attempt at finding a linear model for the collected testing data. The y values of this equation would be exam scores, while the x values would represent hours spent studying. Let us graph this model with our collected data.

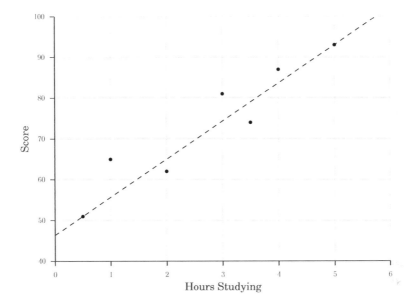

FIGURE 2. First linear model attempt

The linear model does a reasonable job representing the overall trend of the data. That having been said, it's only natural to wonder if there might be a linear model that does a better job? Is there a linear model that would do the best job? To answer this question, we need to quantify exactly what we mean by the linear model "doing a good job" representing the observed data.

Consider the data point $(1, 65)$, which corresponds to 1 hour of study time resulting in an exam score of 65. According to our linear model, one hour of studying should have resulted in an exam score of

$$y = \frac{28}{3}(1) + \frac{139}{3}$$

$$= \frac{28}{3} + \frac{139}{3}$$

$$= \frac{167}{3}$$

$$\approx 55.666$$

so the linear model disagrees with our observed data. Disagreement between a linear model and the observed data is called a *residual*. A residual is defined as

$$\text{Residual} = \text{Observed data value} - \text{Expected data value}$$

where the expected data value is the one predicted by the linear model. Every observed data point will have a residual. For example, the residual for the data point we have been considering, $(1, 65)$ would be

$$65 - 55.666 = 9.333$$

The residuals for each data point are shown below.

Hours Studying	Score	Expected Score	Residual
3	81	$\frac{28}{3}(3) + \frac{139}{3} = 74.333$	$81 - 74.333 = 6.666$
1	65	$\frac{28}{3}(1) + \frac{139}{3} = 56.666$	$65 - 56.67 = 9.333$
2	62	$\frac{28}{3}(2) + \frac{139}{3} = 65$	$62 - 65 = -3$
5	93	$\frac{28}{3}(5) + \frac{139}{3} = 93$	$93 - 93 = 0$
4	87	$\frac{28}{3}(4) + \frac{139}{3} = 83.666$	$87 - 83.666 = 3.333$
3.5	74	$\frac{28}{3}(3.5) + \frac{139}{3} = 79$	$74 - 79 = -5$
0.5	51	$\frac{28}{3}(0.5) + \frac{139}{3} = 51$	$51 - 51 = 0$

Visually, a residual is the vertical distance between the linear model and the observed data. This residual for the data point $(1, 65)$ is shown below in Figure 3.

Residuals represent the error between a linear model and the observed data. The further the observed data is from the linear model, the more error. A linear model that "does the best job" representing the observed data would be the model that minimizes overall model error.

4.1. THE LEAST SQUARES REGRESSION LINE

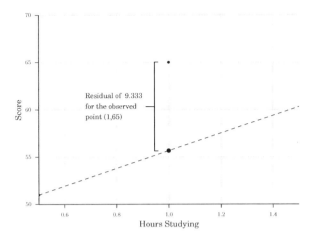

FIGURE 3. Residual for the data point $(1, 65)$.

The total error of our linear model would be all of the individual errors (residuals) added together. However, as some residuals are negative and some are positive, adding the residuals together will not provide us with an accurate accounting of the total error in our model. The positive and negative residuals would negate one another. There are two potential ways we can correct this problem. One would be to take the absolute value of all residuals, the other would be to square all residuals. Both approaches are utilized in statistics, and both yield a meaningful result. We will stick to the more common approach, and square the residuals. Squaring all of the residuals from our earlier calculations would give the following.

Hours Studying	Score	Residual	Residual^2
3	81	6.666	$(6.666)^2 = 44.435$
1	65	9.333	$(9.333)^2 = 87.104$
2	62	-3	$(-3)^2 = 9$
5	93	0	$(0)^2 = 0$
4	87	3.333	$(3.333)^2 = 11.108$
3.5	74	-5	$(-5)^2 = 25$
0.5	51	0	$(0)^2 = 0$

Adding all of these squared residuals together would provide us with a meaningful representation of the total squared error of our model.

$$44.435 + 87.104 + 9 + 0 + 11.108 + 25 + 0 = 176.647$$

The linear model that "does the best job" representing our data would be the model that minimizes this sum. Such a model is formally called the *Least Squares Regression Line*. We will simply refer to it as the regression line from now on.

While the derivation of the regression line is beyond the scope of this text, the understanding, and use of it are certainly not. We will not cover the myriad of ways in which particular calculators and computers can compute the regression line. Whoever is teaching your course will have a method ready to show you (if you happen to be reading this text on your own, a quick search online will yield many different options). Regardless of the device used to calculate it, the regression line for our data is given by

$$y = 8.605x + 49.928$$

Remember, this line is "the best" because it minimizes the sum of the squared residuals. Calculating the squared residuals using the regression line gives the following table.

Hours Studying	Score	Expected Score	Residual	Residual^2
3	81	75.743	5.257	$(5.257)^2 = 27.636$
1	65	58.533	6.467	$(6.467)^2 = 41.822$
2	62	67.138	-5.138	$(-5.138)^2 = 26.399$
5	93	92.953	0.047	$(0.047)^2 = 0.002$
4	87	84.348	2.652	$(2.652)^2 = 7.033$
3.5	74	80.045	-6.045	$(-6.045)^2 = 36.548$
0.5	51	54.23	-3.23	$(-3.23)^2 = 10.436$

The sum of the squared residuals for the regression line is then

$$27.636 + 41.822 + 26.399 + 0.002 + 7.033 + 36.548 + 10.436 = 149.876$$

Note that this sum is smaller than the sum for the first linear model we calculated. The sum of the squared residuals for the regression line will be smaller than the sum of the squared residuals for any other line. This is exactly what makes the regression line "the best", and why it is often referred to as "the best fit line".

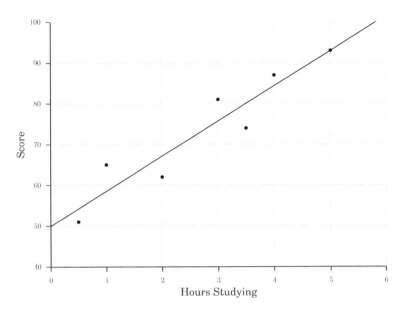

FIGURE 4. The regression line plotted with observed data.

We can use the regression line to predict exam scores based on the amount of time spent studying. As an example, the model predicts that studying 2.5 hours would yield a score of

$$y = 8.605(2.5) + 49.928 \approx 71.44$$

Predictions of mathematical models are split into two types. *Interpolation* is using a value of the independent variable that is between the highest and lowest observed values. In the context of our model, the lowest observed value for our independent variable of time spent studying was 0.5 and the highest was 5. Any predictions we make where the time spent studying is between those two values are interpolations. The example above that predicted a score of 71.44 based on a study time of 2.5 hours was an interpolation.

Extrapolation would result from using a value of the independent variable that is not between the highest and lowest observed values. As an example, for 5.5 hours spent studying, the model predicts an exam score of

$$y = 8.605(5.5) + 49.928 \approx 97.255$$

This prediction is an extrapolation as 5.5 is not in the interval [0.5, 5]. While extrapolation can provide many meaningful results, it is important to realize that mathematical models can only provide a limited amount of meaningful prediction. Consider what happens to this linear model with an input of 12 hours spent studying. The predicted score would be

$$y = 8.605(12) + 49.928 \approx 153.188$$

which no longer makes sense as a real-world prediction. The highest possible score on an exam (barring any extra credit) is 100. This prediction is outside the *scope of the model*. That is to say that the model is no longer capable of providing a meaningful or realistic prediction at this point.

The slope of the regression line provides us with the rate of change of exam scores, based on time studying.

$$m = \frac{\text{change in } y}{\text{change in } x} = \frac{8.605}{1}$$

which would imply that exam scores increase by 8.605 points for every 1 hour that is studied. When the slope of the best fit line is positive, we say that there is a *positive linear relationship* between the independent and dependent variables. If the slope of the best fit line had been negative, we would refer to that as a *negative linear relationship*.

Now that we have "the best" linear model for our data, we can ask the next logical question. How well does this linear model represent the observed data?

In our example of studying and exam scores, the score earned on an exam was dependent on the amount of time spent studying. But, were the exam scores influenced by anything else? Many different factors would contribute to the observed exam scores: difficulty of the exam, blood sugar, stress, the list could go on and on. These unaccounted and unspecified influences are the reason that our linear model has error. If exam scores depended only on the amount of time spent studying, then our data would form a perfectly straight line, and the regression line would pass through every single observed point.

Perfect linear relationships are never observed in real-world scenarios. There will always be outside influential factors that contribute to the observed values of a dependent variable. While there is no way to prevent this unwanted influence, there is a way to quantify it.

The *coefficient of determination*, or r^2, represents the proportion of total variation of the dependent variable that is explained by the regression line. Put another way,

4.1. THE LEAST SQUARES REGRESSION LINE

the coefficient of determination allows us to quantify how much the dependent variable is being influenced by factors other than the independent variable.

The coefficient of determination will always fall on the interval of $[0, 1]$. A value of $r^2 = 0$ would imply that there was no linear relationship, while a value of $r^2 = 1$ would mean that there was a perfect linear relationship. It is important to keep in mind that r^2 being close to 0 does not mean there is no relationship between the independent and dependent variables, rather that there is no linear relationship.

We will fall back on some form of technology to calculate r^2 (as we did for the calculation of the regression line). Regardless of the particular way in which you calculate the coefficient of determination, the value comes out to be

$$r^2 = 0.887$$

This value of 0.887 means that 88.7% of the variation in exam scores is attributed to the amount of time spent studying. Things other than time spent studying accounted for the remaining 11.3% of variation in exam scores. There are no set thresholds that would allow us to say things like "ah, r^2 is above a certain number so our model is good". In the context of this problem, using this simple model, we can feel confident that our model is a good representation of the observed data.

Exercises for Section 4.1.

(1) Pairings for the function f are given below.

x	$f(x)$
2	5
5	6
7	9
10	8
12	12
15	18

(a) Use these pairings to develop a linear regression model for $f(x)$.

(b) Does your linear model do a reasonable job predicting pairings of the function f? Use the coefficient of determination to justify your answer.

(c) Write out a table that lists the residual for each of the points on f.

(d) Add up all of the residuals. Discuss if this sum makes sense.

(e) Use your model to predict the following pairings

(i) (11,   )   (ii) (, 7)

(2) You are working to develop a new type of insulator. The following data represents time since you removed the insulator from the heat source, in seconds, vs temperature, in °C.

Seconds	5	9	11	14	19	25	30
°C	29	28	23	25	16	11	10

Develop a linear regression model for temperature based on an input of seconds. Using the coefficient of determination, discuss the validity of your model. Use your model to predict the temperature of your insulator at 10, 12, and 20 seconds. Finally, determine what might be a realistic scope of your model (this will require some logical reasoning, don't just throw out random numbers).

(3) Vitamin D is produced when the skin on your body is exposed to ultraviolet light. Suppose you perform some data collection where you record sunlight exposure, in minutes, vs measured vitamin D production, in IU (international units). Your recorded data is given in the table below.

Minutes	0.5	1.2	2.3	3	4	6	9	10
IU	87	112	175	256	400	534	810	1,100

Use the least squares regression line to analyze your collected data. Discuss your results.

4.2. Maximizing and Minimizing

Mathematical modeling can be used to answer many different types of questions. The linear model that we developed in the previous section allowed us to predict exam scores based on time spent studying. In this section, we will use models to help us determine optimal solutions to mathematical models in some real-world scenarios. As the upcoming examples will illustrate, optimal solutions occur at maximum or minimum output values.

We will begin with a classic box building example. Suppose you are given a sheet of cardboard that measures 30 inches in length by 12 inches in width. You would like to make a box (without a top) out of this single sheet of cardboard. To do so, you will cut squares out of the edges of the cardboard sheet (illustrated in the figure below), and then fold the resulting sides upward. How large should the squares be (i.e. how far should you cut into the cardboard sheet) so that the volume of the resulting box is maximized?

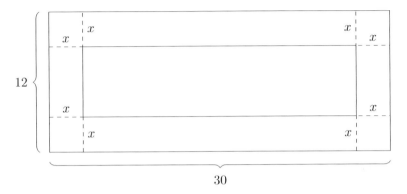

FIGURE 5. The cardboard will be cut along the dotted lines, and folded along the solid lines.

When solving applied problems, it is often helpful to write out a few lines that summarize your goal, as well as the information you are provided with.

- Goal: Maximize volume of the box
- Length of cardboard is 30 inches
- Width of cardboard is 12 inches
- Length of the sides of the squares that will be cut out of the corners is represented by x

Expressing what your variable represents is a simple, but critical part of any applied problem. It helps to clarify the problem setup and any final answers.

The goal is to maximize the volume of a rectangular box. Volume of a rectangular solid is given by

$$V = lwh$$

where l, w, and h represent length, width and height. The lengths of the sides of our box are going to be dependent on the size of the squares we cut out of the corners of the cardboard sheet. Specifically, the length of our box will be $30 - 2x$. The cardboard sheet is 30 inches long, and on each side we are removing x inches. Similarly, the width will be $12 - 2x$. When the sides are folded up, the height of our box will be x.

Plugging these values for length, width, and height into the volume equations yields

$$V = (30 - 2x)(12 - 2x)(x)$$

and we have a mathematical model that will calculate the volume of the box that results from a cut length of x inches. It is entirely reasonable to think of x as an input to this equation, and the volume that results from any particular value of x as output. Furthermore, this equation satisfies the definition of a function. With that in mind, we can write the volume equation in function notation.

$$V(x) = (30 - 2x)(12 - 2x)(x)$$

As a mathematical function, V has a domain of $\mathbb{R}$. However, as an applied model, V does not have a domain of R. The input variable, x, represents the length of the side of a square that will be cut out of the cardboard. Thus, $x > 0$ as it is impossible to have a negative cut length, and a cut length of 0 would leave us with a flat sheet of cardboard. Furthermore, as the width of the cardboard sheet is $(12 - 2x)$ that must mean that $x < 6$, otherwise the cut has gone all of the way through the sheet, $(12 - 2 \cdot 6) = 0$. These real-world restrictions on the input variable mean that as a mathematical model representing volume of a box, the domain of V is $\{x \mid x \in \mathbb{R} \text{ and } 0 < x < 6\}$.

Much like the domain, the range of V is modified if we consider V as a model representing real-world values. $V(x)$ represents the volume of a box. This must mean that $V(x) > 0$. It is not immediately apparent what maximum value the range will achieve (in fact, if we knew that we would be done with this problem). Noting that the range must be positive is enough for the time being.

Now that we have a model that calculates box volume, and we know the domain on which the model has real world meaning, how do we determine the value of x (cut length) that will provide the largest possible box volume? Answering that question algebraically requires a derivative, which is one of the first topics of calculus. You may be wondering why we are covering topics that require calculus in a precalculus text. Algebraically solving maximization or minimization problems for their maximum or minimum values is surprisingly simple (once you are comfortable taking derivatives). The main difficulty in these types of problems is the initial setup, translating the text of the problem into a mathematical model. Translation from text-to-model does not require any topics of calculus (in fact, you might notice that the only math involved in the setup of this problem was the volume of a rectangular solid and some subtraction), and is good preparation for future calculus problems. For the time being, we will once again fall back on some form of technology to calculate the answer.

To begin, graph the model over the real-world domain. Notice that the volume does reach a maximum of approximately 430 in^3 for a cut length between 2 and 3

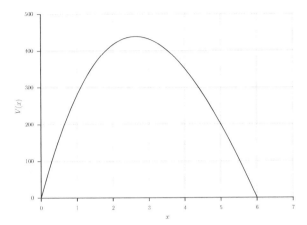

FIGURE 6. Box volume, $V(x)$, based on cut length, x.

inches. Calculating the exact value of this maxima (via a method that will be detailed by your instructor) yields values of $x \approx 2.641$ and $V(x) \approx 438.552$. As x represents cut length and $V(x)$ box volume, a cut length of 2.641 inches would result in a maximum possible box volume of 438.552 in^3.

As another example, suppose you have moved away from a career in box building and into the exciting world of beekeeping. You currently have 40 beehives, each of which produces 33 lbs of honey every season. You would like to add some additional beehives to your collection. For every hive you add, the honey production of the newly added hives will drop by 1 lb per hive (due to things like crowding, competition, disease, etc.). How many hives should you add to maximize honey production?

- Goal: Maximize honey production.
- Currently have 40 hives, each of which produces 33 lbs of honey.
- Every additional hive that is added will reduce the production of all added hives by 1 lb.

To begin, consider how much honey is being produced before any hives are added. If there are 40 hives, and each hive produces 33 lbs of honey, then the total honey production is

$$(40)(33) = 1,320 \text{ lbs}$$

Now, suppose you add one hive. This added hive will produce $33 - 1 = 32$ lbs of honey. If you added two hives, they would each produce $33 - 2 = 31$ lbs of honey, for a total increase of $(2)(31) = 62$ lbs of honey. This pattern will continue, and we can generalize it as

$$\text{Additional honey production} = (x)(33-x)$$

where x represents the added number of hives. The total honey production is this increase, added to the original $(40)(33) = 1,320$ lbs.

$$\text{Total honey production} = (40)(33) + (x)(33-x)$$

In function notation, this would be

$$T(x) = (40)(33) + (x)(33-x)$$

where the input x represents the added number of hives, and $T(x)$ is total honey production. As a function, the domain of T is $\mathbb{R}$. However, as a real-world mathematical model, we must have that $0 \leq x \leq 33$. That is to say we can add somewhere between 0 and 33 beehives to the existing 40. Any other input values prevent this model from representing a plausible real-world quantity.

Some algebraic manipulation shows that

$$T(x) = (40)(33) + (x)(33-x) = -x^2 + 33x + 1320$$

and we can see that the graph of this function is going to be a downward opening parabola. This should make sense intuitively, as we are expecting this function to achieve a maximum value somewhere on the real world domain of $0 \leq x \leq 33$, and a downward opening parabola will have a maximum value at its vertex. Figure 7 illustrates the graph of T on its real-world domain.

While we could fall back on some form of technology to calculate the maximum total honey production, we can also answer this particular question algebraically. As previously mentioned, the maximum vertical value this function achieves would be the vertex of the parabola. The x coordinate of the vertex is

$$x = \frac{-b}{2a} = \frac{-33}{2(-1)} = 16.5$$

which would suggest that adding 16.5 beehives would result in maximum honey production. Once again, we find a distinction between theoretical and real-world values. Adding half of a beehive is not possible, so we need to decide if we should round down to 16 hives, or up to 17 hives. Comparing the total honey production for 16 and 17 hives provides an interesting realization.

$$T(16) = -(16)^2 + 33(16) + 1320 = 1592 \text{ lbs}$$

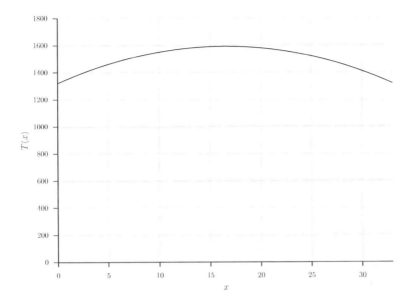

FIGURE 7. Added beehives, x, plotted with total honey production, $T(x)$.

$$T(17) = -(17)^2 + 33(17) + 1320 = 1592 \text{ lbs}$$

The total honey production for 16 hives is the same for the total honey production for 17 hives. Why? The symmetry of a parabola. The x coordinate of the parabola's vertex was 16.5. All parabolas are symmetric across the vertical line passing through their vertex (in this case the line is given by the equation $x = 16.5$). Because the values of 16 and 17 are both the same horizontal distance away from 16.5 (they are both a distance of 0.5 away from 16.5), the vertical values of the parabola must be the same for these x coordinates.

Concerning our initial goal, maximizing total honey production, it makes more sense to choose the x value of 16. As the x variable represents additional beehives that will be added to the original 33, there is no reason to add 17 hives when adding 16 results in the same total honey production. In summary, you should add 16 hives to the existing 33, for a total of 49 hives that together will produce $1,592$ pounds of honey.

For a final example, we want to produce a can that must hold 250 cubic centimeters of liquid. We know that the material to make the can costs $0.05 per square centimeter. What are the dimensions of the can that result in a minimum cost of production?

- Goal: Minimize cost of the can
- Volume of the can is 250 cm^3

- Cost is $0.05 per cm^2

The information about the required volume allows us to write out the volume equation of a cylindrical solid.

$$V = \pi r^2 h = 250$$

Note that it is not immediately clear how this relationship will be applied to our stated goal of minimizing the cost of the can. We are simply organizing and clarifying all of the provided information.

The cost of the material has units of dollars per square centimeter. To find the total cost of the can, we would need to multiply the cost per square centimeter by the total number of square centimeters required to construct the can. In other words, we need to determine the surface area of the can.

Unlike volume, the surface area of a cylindrical solid may be something new to you. Nevertheless, it is well within our grasp. Imagine taking a hollow cylindrical can, and cutting off the top and bottom. Now, cut vertically through the remaining hollow tube and unroll the tube. You have three known shapes: Two circles (the top and bottom), and a rectangle. Each of the top and bottom have area πr^2. The rectangle has dimensions h, the height of the original can, and $2\pi r$, the circumference of the top (or bottom) circle. This means the area of the sides would be given by $2\pi r h$. Thus, the surface area of the total can would be

$$\text{Surface Area} = 2\pi r^2 + 2\pi r h$$

As was previously discussed, the total cost for building the can would be the cost per square centimeter multiplied by the total number of square centimeters. Mathematically, we would have

$$C = 0.05 \left(2\pi r^2 + 2\pi r h\right)$$

Our goal is to minimize cost, and we have arrived at the cost equation for can production. There is, however, a problem. The cost equation contains two unknown values, the radius, r, and height, h. We cannot minimize an equation with two variables.

Recall that the first piece of information we organized was the volume relationship. This relationship is an equation with the same two unknown variables. In other words, we have a system of equations. We can solve the volume equation for either of the variables, and substitute that value into the cost equation. Solving the volume equation

for h would give
$$h = \frac{250}{\pi r^2}$$
and substituting this into the cost equation would leave us with
$$C = 0.05\left(2\pi r^2 + 2\pi r\left(\frac{250}{\pi r^2}\right)\right)$$

The cost function now contains a single variable, r. We can write the cost as a function of can radius and simplify
$$C(r) = 0.05\left(2\pi r^2 + \frac{500}{r}\right)$$

Before using technology to calculate the minimum, we should consider real-world domain and range values. As the input represents the radius of a can, $r > 0$ seems to make good sense. It is not immediately clear what a maximum value for r may be, so we estimate something reasonable, $r < 10$. Similarly, as the range represents cost, $C(r) > 0$.

On the real-world domain, the cost function achieves a minimum at the point $(3.414, 10.984)$. This tells us that the minimum cost of the can is approximately \$10.98, and that the minimum cost is achieved when the can radius is roughly 3.414 cm.

As illustrated by the examples thus far, all applied problems are different. No algorithm will solve all the word problems you may encounter, and you should avoid trying to develop any such mentality, as there will always be situations where an algorithm will fail. However, there are some thoughts, and habits that will help with approaching, understanding, and solving all applied problems.

- List the information you know, and the question(s) you would like to answer.

- If applicable, draw a picture of the scenario.

- Clearly state the quantity that your variable represents.

- Consider relationships between the information you know, and the information you would like to know.

- Go slowly, ask yourself questions, and don't get discouraged if the answers are not immediately apparent.

Exercises for Section 4.2.

(1) Identify the maximum and minimum values of the following functions on the interval $[0, 25]$.

(a) $f(x) = \dfrac{1}{9}x^2 - \dfrac{10}{3}x + 31$

(b) $g(x) = \dfrac{\left(\frac{1}{7}x - 1\right)^3 + 2}{x}$

(2) Suppose you build a cannon. After careful observation, you determine that the height of cannon balls fired from your cannon can be modeled by the function
$$h(t) = -9.8t^2 + 75t + 1$$
where t is time in seconds, and $h(t)$ is meters above the ground.

(a) Explain $h(0)$ in real world terms. Does it make sense?

(b) What is the real world domain of h?

(c) What is the maximum height of a cannon ball? How long after being fired does the ball reach this height?

(3) While attending college you try to work out a balance between your study habits and social life. Starting from a value of 0.4, your G.P.A. will increase by 0.35 points for every 8 hours that you spend studying during a week. At the same time, every 8 hours that you spend studying during a week will decrease your social circle of "friends" by 7% (assume you start with 100% of a social circle). You want to maximize your life, which is the product of studying and socializing. How long should you spend studying each week?

(4) Consider the graph of the fundamental function
$$f(x) = \sqrt{x}$$

(a) What point on the graph of f is the closest to the point $(3, 0)$?

(b) Suppose you want to draw a rectangle that must be under the graph of f, and to the left of the vertical line $x = 12$. The rectangle may touch the graph of f, and the vertical line of $x = 12$. What are the dimensions of the rectangle that result in the largest possible area?

(5) You have been hired to construct a Norman window. Norman windows consist of a half circle on top of a rectangular window. If the perimeter of the window will be 7 meters, what dimensions of the window will result in it having the largest possible area?

(6) After a long day of walking on the beach, you decide it's time to head back to your car. You are 2 kilometers south of the road that will lead you back to your car. All of the terrain between you and the road (in all possible directions) is soft beach sand. Once on the road, your car is an additional 3 kilometers to the east. Your walking speed on the beach sand is 1.4 kilometers per hour, and your walking speed on the road is 3 kilometers per hour. What path should you walk to minimize the time it takes you to return to your car?

(7) A local amusement park charges $12 for a single admission. They will discount all tickets in a group by $0.50 for each person over a group size of 20.

   (a) Write a function that models the cost of a ticket, based on an input of group size. You may assume that the group size is 20 or more.

   (b) Write a function modeling the amusement park revenue based on an input of group size. You may assume that the group size is 20 or more.

   (c) What group size results in the maximum amount of revenue for the amusement park?

(8) You and two of your friends are trying to connect your houses to a nearby fiber optic cable. The three of you must all connect to the cable at the same point, and then branch to each of your individual houses. Where should you connect to the fiber optic cable so that you minimize the total length of cable that will run to each of your houses? The figure below illustrates the respective distances between your homes and the cable.

(9) Your neighbor owns a farm and has asked for your help in building some new pens for pigs. The pens will consist of one large rectangle, divided into four smaller, equally sized, rectangular pens, each of which has a width equal to the large rectangular pen. Your neighbor has 200 meters of fencing with which to build the new pens. What are the dimensions of the large

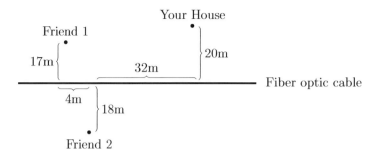

rectangular pen that will maximize the dimensions of the smaller pens? What are the dimensions of the smaller pens?

(10) A series of life choices has found you working as an engineer at an energy drink company. Your current task is to create a new can that must have a volume of 350 cubic centimeters. Per square centimeter, the material for the top, sides, and bottom costs

$$\text{Top}: \$0.15$$
$$\text{Sides}: \$0.08$$
$$\text{Bottom}: \$0.10$$

What are the dimensions of the can that minimize the production cost?

4.3. Exponential Growth and Decay

4.3.1. Simple Interest. Our goal in this section is to model how a quantity increases or decreases by a fixed percent at regular intervals. As an example, suppose you deposit $100 into a bank account that pays 6% interest, compounded annually (we also suppose that you will not make any other deposits or withdraws). The phrase "6% interest compounded annually" means that at the end of every year the bank adds 6% of your account value to your account. At the end of the first year the account total would be

$$\underbrace{100}_{\text{Starting } \$100} + \overbrace{(100)(0.06)}^{6\% \text{ of } \$100} = \$106$$

To calculate your account value at the end of the second year, you would not add another $6 to your account total. Instead, you would add 6% of your account value which is now $106.

$$106 + (0.06)(106) = \$112.36$$

The account value at the end of the second year was dependent on the account total, which had been increased by the interest from the first year. This idea becomes more apparent if we write out several years of interest payments all at once (note that the equations below are the same as the ones above, but they have been factored).

First year:	$100(1 + 0.06) = 106$
Second year:	$106(1 + 0.06) = 112.36$
Third year:	$112.36(1 + 0.06) = 119.1016$
Fourth year:	$119.1016(1 + 0.06) = 126.2476...$

Each successive year depends on the one before it. This is a pattern that we can generalize.

Rather than pick a specific starting amount of money, we use a variable, A_0, to represent any starting amount we might want. Additionally, A_n will represent the account value at the end of year n. We will continue to use 6% interest for the time being. At the end of the first year, the account value would be

$$A_1 = A_0(1 + 0.06)$$

Following the pattern from above, the account value at the end of the second year would be

$$A_2 = A_1(1 + 0.06)$$

We can simplify this equation for A_2 by substituting $A_1 = A_0(1 + 0.06)$ in for A_1

$$A_2 = A_1(1 + 0.06) \quad \text{substitute in for } A_1$$
$$= A_0(1 + 0.06)(1 + 0.06)$$
$$= A_0(1 + 0.06)^2$$

Continuing on for the account balances at the end of year three and four

$$A_3 = A_2(1 + 0.06) \quad \text{substitute in for } A_2$$
$$= A_0(1 + 0.06)^2(1 + 0.06)$$
$$= A_0(1 + 0.06)^3$$

$$A_4 = A_3(1 + 0.06) \quad \text{substitute in for } A_3$$
$$= A_0(1 + 0.06)^3(1 + 0.06)$$
$$= A_0(1 + 0.06)^4$$

This pattern will continue, and we can generalize it as the following

$$A_n = A_0(1 + 0.06)^n$$

where n is any year of our choosing. Rewriting A_n as $A(n)$ allows us to express this equation in function notation.

$$A(n) = A_0(1 + 0.06)^n$$

This function takes an input of n years, and gives account value at the end of year n as output. As an example, with a starting deposit of $A_0 = \$5,000$ after 15 years the account value would be

$$A(15) = 5,000(1 + 0.06)^{15} = \$11,982.79$$

There is no reason that we need to restrict this equation to an interest of 6% (we stuck with it up until now only for purposes of the example). In the same way we generalized the starting amount as A_0, we can generalize the interest as I. Doing so produces our first model for exponential growth.

$$A(n) = A_0(1 + I)^n$$

This equation models exponential growth and decay for a quantity that grows or decays by a fixed percent at regular intervals. A_0 is the initial starting amount, n is the number of time intervals that have passed (it need not be years), and I is the growth or decay rate. If $I > 0$ it is referred to as a growth rate. If $I < 0$ it is referred to as a decay rate.

4.3.2. Compound Interest. Our work in the previous section allowed us to model how a quantity increased or decreased if it was changed by a fixed percent at regular intervals. The idea for this section is slightly different. Suppose, once again, we have a bank account that pays us yearly interest. This time, we receive 100% yearly interest, but with quarterly compounding. Quarterly compounding would mean that the total interest will be split into four different compounding events. In the context of this problem, that would mean that at four different points throughout the year, we would receive an interest payment of 25%. Together, the interest of these four compounding events adds to our total yearly interest of 100%.

According to our previous exponential growth model, the four compounding events of the first year would be

$$A(1) = A_0(1 + 0.25)$$

$$A(2) = A_0(1 + 0.25)^2$$

$$A(3) = A_0(1 + 0.25)^3$$

$$A(4) = A_0(1 + 0.25)^4$$

where $A(4) = A_0(1 + 0.25)^4$ would represent the total amount of money in the account at the end of the first year. Rewriting the decimal of 0.25 as a fraction would leave us with

$$A(4) = A_0 \left(1 + \frac{1}{4}\right)^4$$

where the 1 in the numerator represents the total interest payment (in this case 100%), and the 4 in the denominator represents the number of times the total interest is being "split" into separate compounding events.

The difference between quarterly compounding and yearly compounding can be substantial. A comparison of one year at 100% interest with quarterly compounding, and one year at 100% interest with yearly compounding is shown below.

Quarterly Compounding	Yearly Compounding
$A_0 \left(1 + \frac{1}{4}\right)^4 = 2.44 A_0$	$A_0(1 + 1) = 2A_0$

At the end of the year, a quarterly compounded account would have 2.44 times the starting value, while a yearly compounded account would merely double.

Suppose we want to compare these two accounts after two years of interest. Every year of quarterly compounding involves 4 compounding events, each of which is 25%. Over the course of two years, there would be $4 \cdot 2 = 8$ of these events.

Quarterly Compounding	Yearly Compounding
$A_0 \left(1 + \frac{1}{4}\right)^{4 \cdot 2} = 5.96 A_0$	$A_0(1 + 1)^2 = 4A_0$

Once again, we can generalize this idea. Generalizing to an interest of I, and k compounding events will provide us with an exponential model that calculates compound interest.

$$A(N) = A_0 \left(1 + \frac{I}{k}\right)^{kN}$$

In this model, $A(N)$ is the account value after N years, I is the yearly interest rate, and k is the number of yearly compounding events. Note that if $k = 1$ then this model reduces to the simple interest model from the previous section.

As an example, suppose we have \$5,000 that we deposit into an account that pays 1.75% yearly interest, compounded monthly. What would the account value be after 15 years?

$$A(15) = 5{,}000 \left(1 + \frac{0.0175}{12}\right)^{12 \cdot 15} = \$6{,}499.64$$

At this point, our model is complete. There is, however, an interesting idea looming just beneath the surface of our model. Let us consider the quantity

$$\left(1 + \frac{1}{k}\right)^k$$

In the context of our model, this would represent an interest rate of 100% being split into k compounding events. As we increase k, what happens to this quantity?

k	$\left(1 + \frac{1}{k}\right)^k$
1	2
2	2.25
5	2.48832
20	2.65329...
50	2.69158...
500	2.71556...
5000	2.71801...
50000	2.71825...

This table suggests that the quantity $\left(1 + \frac{1}{k}\right)^k$ increases less and less for larger values of k. If we take this idea to the extreme and let k increase without bound (another way to say that is we let k go to infinity), what happens? In fact, there is a boundary, or a *limit* that the quantity $\left(1 + \frac{1}{k}\right)^k$ will approach. We define this as

$$\lim_{k \to \infty} \left(1 + \frac{1}{k}\right)^k = e$$

The above line would be read as "the limit as k goes to infinity of $\left(1 + \frac{1}{k}\right)^k$ is e". This means that as k increases, the quantity $\left(1 + \frac{1}{k}\right)^k$ will continue to get closer to, but never go beyond, the number e. How close will it get? As close as we want it to (this idea will be one of the first topics of calculus).

Although we defined the number e through the use of a limit, e is just a real number. More specifically, it is an irrational number. That is to say that e is a decimal number that goes on forever and never repeats. The first few decimal digits are $e = 2.7182...$

In the context of our financial problem, e would represent continuous compounding (compounding at all possible opportunities) at 100% yearly interest ($I = 1$). Replacing $\left(1 + \frac{1}{k}\right)^k$ with e would give

$$A(N) = \lim_{k \to \infty} A_0 \left(1 + \frac{1}{k}\right)^{kN} = A_0 e^N$$

Compare one year of continuous compounding at 100% interest to the previous examples of yearly, and quarterly compounding.

Yearly	Quarterly	Continuous
$A_0(1+1) = 2A_0$	$A_0\left(1 + \frac{1}{4}\right)^4 = 2.44 A_0$	$A_0 e^1 \approx 2.718 A_0$

Continuous compounding results in the largest increase as you would literally be compounding your interest at all possible opportunities.

While this discussion of limits, e, and continuous compounding is interesting, you may be wondering how it applies to situations where the interest rate is not 100%. Though it is beyond the scope of this text to prove, it is also true that

$$\lim_{k \to \infty} \left(1 + \frac{I}{k}\right)^k = e^I$$

where I is any real number. In other words, we can use e, and the idea of continuous compounding, for any interest rate we might want.

From our earlier model we have

$$A(N) = A_0 \left(1 + \frac{I}{k}\right)^{kN}$$

If we replace the $\left(1 + \frac{I}{k}\right)^k$ with e^I, we would have a new model based on continuous compounding. Let us call this new model F.

$$F(N) = A_0 e^{IN}$$

Our model for F is complete, but there is one final clarification. In both models we have used the variable I to represent the interest rate (or rate of change). The only instance when both models would use the same value for I is continuous compounding ($k \to \infty$). In every other circumstance, the models would use different values for their respective rates of change (values for I). To prevent confusion, we will use a different letter, r, to represent the rate of change in model F.

To summarize, we have developed two different models for exponential growth and decay. The first is given by

$$A(N) = A_0 \left(1 + \frac{I}{k}\right)^{kN}$$

where A_0 is the starting amount, N is the time interval, I is the rate of change (positive or negative) for a time interval, and k is the number of compounding events per time interval.

The second model is based on continuous compounding, $k \to \infty$, and is given by

$$F(N) = A_0 e^{rN}$$

where A_0 is the starting amount, N is the time interval, and r is referred to as the exponential growth or decay constant.

For now, the vast majority of our exponential modeling will be done with the model we named A. The ideas of e and the model F were presented because they both came naturally out of the discussion. This is not to say that e and the model F are not important, they are. Our future work will involve both of them extensively. In time, we will consider the function $f(x) = e^x$ as a fundamental exponential function, and we will be able to move between the models of A and F effortlessly.

Let us conclude this section with a classic example in exponential modeling. The half-life of carbon-14 is approximately 5,730 years. Put another way, regardless of the amount of carbon-14 you start with, half of it will have radioactively decayed after

4.3. EXPONENTIAL GROWTH AND DECAY

5,730 years. We can use this information to come up with a function that will model the decay of carbon-14 over time.

Starting with our generic model

$$A(N) = A_0 \left(1 + \frac{I}{k}\right)^{kN}$$

we can begin to identify the particular values for our variables. Given that we will lose 50% of our starting amount (regardless of what that amount is) over the course of 5,730 years, it seems that $I = -0.5$ would be a reasonable choice. As this loss occurs once every 5,730 years, a value of $k = 1$ is also correct. Plugging these values into our model would leave us with

$$A(N) = A_0 \left(1 + \frac{-0.5}{1}\right)^{1 \cdot N} = A_0 \left(\frac{1}{2}\right)^N$$

This model is correct, but it has an awkward time unit. The choices of $I = -0.5$ and $k = 1$ were based on a time interval of 1 half-life, or 5,730 years. As written, the N in our model does not represent years, but instead represents half-lifes. $N = 1$ would correspond to 1 half-life, or 5,730 years. $N = 2$ would correspond to 2 half-lifes, or 11,460 years. To convert our model over to a time unit of years (which would be vastly preferable), we need to divide the number of half-lifes by 5,730.

$$A(N) = A_0 \left(\frac{1}{2}\right)^{\frac{N}{5,730}}$$

Now N is in a time interval of years, and our model is complete.

Suppose we had 15 grams of carbon-14. How much would remain after 6,250 years? Letting $A_0 = 15$ and $N = 6,250$ would give

$$A(6,250) = 15 \left(\frac{1}{2}\right)^{\frac{6,250}{5,730}} \approx 7.042$$

Thus, after 6,250 years, only 7.042 grams of carbon-14 would remain.

Exercises for Section 4.3.

(1) You invest $7,500 in an account that pays 3.5% interest, compounded annually. Assume you will make no withdraws on the account.

   (a) What will your account value be at the end of 6 years?

   (b) What will your account value be after 6.5 years?

(c) Compare the first two parts of this question and discuss why your answers make sense.

(2) Many financial institutions offer daily compounding on account balances. Suppose you invest $15,500 into an account that pays 2.1% yearly interest, compounded daily. Assume you will make no withdraws on the account.

   (a) What will your account value be at the end of 4 years?

   (b) What will your account value be at the end of 4.5 years?

   (c) Compare the first two parts of this question and discuss why your answers make sense.

(3) What yearly interest rate would be required for an investment of $8,000 to grow to $12,400 after two years of monthly compounding?

(4) Though we won't go into the details of the derivation, the equation

$$(1+r)^n = \frac{W}{W - Pr}$$

illustrates the amount of money you would be able to withdraw on a yearly basis, W, if you invested P dollars into an account that had an interest rate of r. You would be able to make this withdraw every year for n years. Solve this equation for W.

(5) The half life of fermium-253 is approximately three days. Suppose that while walking in the park, you find a pile of fermium-253. Intrigued, you take the pile home and observe the radioactive decay. How much of the pile will remain in 65 days? Ignore the fact that you have long since died of radiation poisoning.

(6) Your neighbor has recently become the proud owner of a large floodlight and enjoys illuminating your nighttime hours. To rectify this situation, you decide to purchase some darkened glass for your windows. Each centimeter of darkened glass removes 27% the light that passes through it.

   (a) Write a model where the input is centimeters of glass and the output is percent of light that has passed through.

(b) Evaluate a few different thickness choices (up to you), and comment on which you would select.

(7) Suppose you have a single bacteria that is able to double every minute.

   (a) Write a model for the bacterial population growth.

   (b) At 8:00 am, you place this single bacteria into a petri dish. By 8:20 am, the dish is full of bacteria. When was the petri dish half full?

   (c) Suppose that at 8:15 am your bacteria realize their population growth is unsustainable and that something must be done to prevent their societal collapse due to overpopulation. Through good fortune, they discover three nearby, unoccupied, and habitable petri dish environments, each of which is the same size as their original dish. Assuming their growth rate remains the same, how much additional time did this discovery allow them?

   (d) Discuss your results from the previous part with regard to human population growth, and the idea that colonizing additional worlds will solve potential overpopulation problems.

(8) You have developed a new type of pain killer and are proceeding through clinical trials. Your new drug is metabolized by the liver at a rate of 18% per hour. Write a model that provides the concentration of the drug in a person as a function of time. If a patient consumes 450 mg of drug at 11:00 am, how much will remain in their bloodstream at 12:45 pm?

(9) In this exercise you will compare several different exponential models. For each of the models, the yearly interest rate is 3.2%. One of the models experiences monthly compounding, another other weekly, another daily, and the last is continuous. Discuss how the account balances compare after two, five, and fifteen years. Use some graphing software to graph the models. Discuss your results.

CHAPTER 5

Exponential and Logarithmic Functions

This chapter will formally develop the topics of exponential and logarithmic functions. To that end, the number e was defined in a previous chapter. If you are not yet comfortable with the number e, you should take some time and revisit its definition.

5.1. Exponential Functions

The general form of an exponential function is given by

$$f(x) = b^x$$

where the base, b, is a real number strictly greater than zero. The restriction that $b > 0$ helps avoid potential problems that would arise if b was less than zero, and x was rational with an even denominator (i.e. $(-1)^{\frac{1}{2}}$ is not a real number).

For the function $f(x) = b^x$ the following three pairings

x	$f(x)$
-1	$\frac{1}{b}$
0	1
1	b

are a good starting point when graphing or transforming f.

As an example, consider these pairings by the functions $g(x) = 2^x$ and $h(x) = e^x$

x	$g(x)$		x	$h(x)$
-1	$\frac{1}{2}$		-1	$\frac{1}{e}$
0	1		0	1
1	2		1	e

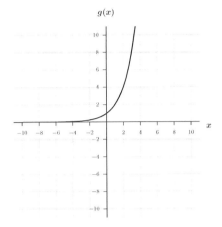

FIGURE 1. $g(x) = 2^x$

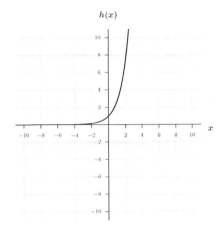

FIGURE 2. $h(x) = e^x$

Note the similarities between the graph of g (Figure 1) and the graph of h (Figure 2): both are increasing; both pass through the point $(0, 1)$; both have domain $\mathbb{R}$ and range strictly greater than zero. Additionally, both functions have the x-axis as a *horizontal asymptote*. We mention this fact only as horizontal asymptotes are an important characteristic of exponential functions. A full discussion of horizontal asymptotes is reserved for a later section.

The properties of exponents play a large role in exponential functions. Consider the functions

$$g(x) = 3^{-x} \quad \text{and} \quad h(x) = \left(\frac{1}{3}\right)^x$$

A quick check of the three specific input values $-1, 0, 1$ that were mentioned earlier shows the following pairings.

x	$g(x)$
-1	3
0	1
1	$\frac{1}{3}$

x	$h(x)$
-1	3
0	1
1	$\frac{1}{3}$

While this does not prove that the functions g and h are equal, it definitely suggests as much. Using the laws of exponents, we can manipulate 3^{-x} as follows

$$3^{-x} = 3^{-1 \cdot x} = (3^{-1})^x = \left(\frac{1}{3}\right)^x$$

and we have shown that the equations responsible for the pairings of g and h are algebraically equivalent. As all respective pairings of g and h are identical, the functions are equal.

The graph of g (Figure 3) illustrates the notion of *exponential decrease*. Exponential decrease occurs when the base of an exponential function is between zero and one, or when there is a negative sign on the exponent.

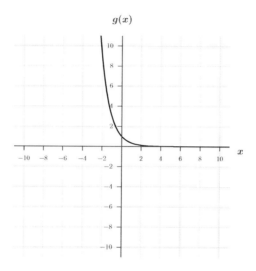

FIGURE 3. $g(x) = \left(\frac{1}{3}\right)^x$

One final interpretation for the function $g(x) = 3^{-x}$ would be to consider g as a transformation on the function $f(x) = 3^x$. Specifically, g would be a horizontal reflection across the vertical axis. Graph the function $f(x) = 3^x$ and verify this transformation for yourself.

Exercises for Section 5.1.

(1) Sketch a graph for each of the following exponential functions. Draw these graphs by hand, do not use graphing software.

(a) $f(x) = -3^{x-4}$ (b) $g(x) = 5^{-x} + 1$ (c) $h(x) = \dfrac{1}{2} \cdot 2^{x+1}$

(2) Each of the following functions are some number of transformations on a general exponential function. State the general exponential, and the corresponding points of $(-1, \frac{1}{b})$ and $(1, b)$. Then, list the transformations that result in the given function and perform the required transformations

on the points $(-1, \frac{1}{b})$ and $(1, b)$. Verify that your transformed points satisfy the given function.

(a) $j(x) = -2^{x+7}$

(b) $k(x) = \frac{3}{5}e^{-x}$

(c) $l(x) = \left(\frac{1}{2}\right)^{\frac{1}{2}x+4}$

(d) $m(x) = 4 \cdot 5^{3-x}$

(e) $n(x) = -6 \cdot \left(\frac{9}{5}\right)^{\frac{2}{11}x+1} - 2$

(f) $p(x) = e \cdot e^{-x} - e$

(3) Each of the graphs given below are some number of transformations on a general exponential function. Determine the equation of each function. You may assume plotted points and horizontal asymptotes occur on integer values.

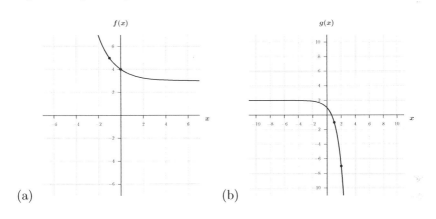

(a) (b)

(4) Use algebra to prove if the following functions make all of the same pairings. You should also state the domains of f and g. Based on your work, can you conclude $f = g$?

$$f(x) = 2^{2x-1} \quad \text{and} \quad g(x) = \frac{1}{2}4^x$$

(5) Consider the functions

$$f(x) = \frac{e^x - e^{-x}}{2} \quad \text{and} \quad g(x) = \frac{e^x + e^{-x}}{2}$$

Prove which of these functions is even and which is odd.

5.2. Logarithmic Functions

As exponential functions are bijective, they must be invertible. For a general exponential function of the form

$$f(x) = b^x$$

the domain values (the exponents) are paired with the base raised to that exponent. The inverse function would make these pairings in exactly the opposite order. That is to say that the inverse function would have a domain of a particular base raised to an exponent, paired with the exponent.

$$\text{Exponent} \underset{f^{-1}}{\overset{f}{\rightleftarrows}} \text{Base raised to exponent}$$

The three pairings of $f(x) = b^x$ mentioned earlier

x	$f(x)$
-1	$\frac{1}{b}$
0	1
1	b

would become the following pairings in the inverse function

x	$f^{-1}(x)$
$\frac{1}{b}$	-1
1	0
b	1

The general idea behind the pairings of the inverse function goes as follows: Given a domain value, what exponent on the base caused this result? Consider the pairing of $(\frac{1}{b}, -1)$. The inverse function makes this pairing because $\frac{1}{b}$ is the result of an exponent of -1 on base b. This idea is how we arrive at the definition of a *logarithm*.

DEFINITION 5.1. Logarithm

For an exponential equation of the form $f(x) = b^x$, where $b > 0$, the inverse of f is given by $f^{-1}(x) = \log_b x$.

5.2. LOGARITHMIC FUNCTIONS

Let us illustrate this definition with some examples.

$$\log_2 8 = 3 \quad \text{because} \quad 2^3 = 8$$
$$\log_5 25 = 2 \quad \text{because} \quad 5^2 = 25$$
$$\log_2 \tfrac{1}{8} = -3 \quad \text{because} \quad 2^{-3} = \tfrac{1}{8}$$

It is common practice to write any logarithm involving a base of 10 without a base. Rather than write $\log_{10} 100 = 2$ we would write $\log 100 = 2$. Whenever you come across a logarithm without a base, assume the base is 10.

Similarly, a base of e is such a common occurrence that instead of writing $\log_e$ we simply write ln. As an example, we would not write $\log_e e^3 = 3$, but would instead write $\ln e^3 = 3$. Whenever a logarithm involves base e we refer to it as the *natural logarithm*.

Our previous work with inverse functions tells us that the domain of a logarithmic function must be the range of the exponential, $\{x \mid x \in \mathbb{R} \text{ and } x > 0\}$, and that the range of a logarithmic function must be the domain of the exponential, $\{f^{-1}(x) \mid f^{-1}(x) \in \mathbb{R}\}$.

There are several properties of logarithms that will be extremely useful to us. Given the functions

$$f(x) = b^x \quad \text{and} \quad f^{-1}(x) = \log_b x$$

we know that the compositions

$$f^{-1}(f(x)) = \log_b(b^x) = x$$

on the domain of f, and that

$$f(f^{-1}(x)) = b^{\log_b x} = x$$

on the domain of f^{-1}. These composition relationships are extremely useful algebraic tools.

Though we will not prove them here, the following theorems will also play a large role in upcoming algebraic manipulation. For $b > 0$ and $b \neq 1$

(1) $b^x = b^y$ if and only if $x = y$
(2) $\log_b x = \log_b y$ if and only if $x = y$

Logarithms are exponents. This fact means that the product, power, and quotient laws of exponents will all have a logarithmic version. Recall the product rule for exponents.

$$b^r \cdot b^s = b^{r+s}$$

Now, consider the expression

$$b^{\log_b(rs)}$$

According to the composition relationships from above, it must be true that

$$b^{\log_b(rs)} = rs$$

It is also true that

$$b^{\log_b(rs)} = rs = \left(b^{\log_b r}\right)\left(b^{\log_b s}\right)$$

From the product rule of exponents we have

$$b^{\log_b(rs)} = rs = \left(b^{\log_b r}\right)\left(b^{\log_b s}\right) = b^{\log_b r + \log_b s}$$

We have arrived at the relationship that $b^{\log_b(rs)} = b^{\log_b r + \log_b s}$. As the single base on both sides of this equation is the same, it must be true that the exponents are equal (from the theorems given above). This fact leaves us with the *product rule for logarithms*.

$$\log_b(rs) = \log_b r + \log_b s$$

Recall the quotient rule for exponents.

$$\frac{b^r}{b^s} = b^{r-s}$$

Consider the expression

$$b^{\log_b \frac{r}{s}}$$

Following a similar series of steps, we can arrive at the relationship

$$b^{\log_b\left(\frac{r}{s}\right)} = \frac{r}{s} = \frac{b^{\log_b r}}{b^{\log_b s}} = b^{\log_b r - \log_b s}$$

from which we have the *quotient rule for logarithms*

$$\log_b\left(\frac{r}{s}\right) = \log_b r - \log_b s$$

Lastly, recall the power rule for exponents.

$$(b^r)^s = b^{rs}$$

Suppose we have that $\log_b a = d$. The definition of a logarithm tells us that $b^d = a$, from which we can conclude

$$\left(b^d\right)^c = a^c$$

$$b^{cd} = a^c$$

This tells us that the exponent we need to put on b to get a^c is cd. Or, in logarithmic form, $\log_b(a^c) = cd$. Substitution in with $d = \log_b a$ yields the *power rule for logarithms*.

$$\log_b(a^c) = c \log_b a$$

The definition, properties, and rules of logarithms are now at our disposal. Let us illustrate their use on some examples.

Simplify the expression $\log_4 [\log_3 (\log_2 8)]$

$\log_4 [\log_3 (\log_2 8)]$	Use the definition of a logarithm on $\log_2 8$
$\log_4 [\log_3 3]$	Use the definition of a logarithm on $\log_3 3$
$\log_4 1$	Use the definition of a logarithm on $\log_4 1$
0	Final simplified form

Expand the expression $\log_2 \left(\dfrac{x^4}{\sqrt{y} \cdot z^3}\right)$

$$\log_2\left(\frac{x^4}{\sqrt{y}\cdot z^3}\right) = \log_2 x^4 - \log_2\left(\sqrt{y}\cdot z^3\right)$$

$$= \log_2 x^4 - \left(\log_2 y^{\frac{1}{2}} + \log_2 z^3\right)$$

$$= 4\log_2 x - \frac{1}{2}\log_2 y - 3\log_2 z$$

Write the following as a single logarithm

$$2\log_{18} a + 3\log_{18} b - \frac{1}{3}\log_{18} c = \log_{18} a^2 + \log_{18} b^3 - \log_{18} c^{\frac{1}{3}}$$

$$= \log_{18}\left(a^2 b^3\right) - \log_{18} c^{\frac{1}{3}}$$

$$= \log_{18}\left(\frac{a^2 b^3}{c^{\frac{1}{3}}}\right)$$

5.2.1. Change of Base. You may find yourself working with a logarithmic base that is somewhat inconvenient. As an example, suppose you came across the quantity

$$\log_3 7$$

While there is nothing mathematically wrong with this expression, working in bases 10 and e will have advantages in the future. To that end, we can rewrite this expression using either of those bases.

Suppose we let $y = \log_3 7$. We then get the exponential relationship $3^y = 7$.

$$3^y = 7$$

$$\log(3^y) = \log 7$$

$$y \log 3 = \log 7$$

$$y = \frac{\log 7}{\log 3}$$

These algebraic steps have resulted in an equivalent expression for $\log_3 7$ that uses logarithms of base 10. This idea can be generalized, and is formally known as *change of base*.

If $y = \log_a b$ then $a^y = b$.

$$a^y = b$$

$$\log_c(a^y) = \log_c b$$

$$y \log_c a = \log_c b$$

$$y = \frac{\log_c b}{\log_c a}$$

where c is a base of your choosing.

The following examples illustrate change of base.

(1) $\log_5 13 = \dfrac{\ln 13}{\ln 5}$ where base e was chosen.

(2) $\log_4 3 = \dfrac{\log 3}{\log 4}$ where base 10 was chosen.

(3) $\log_{11} 7 = \dfrac{\ln 7}{\ln 11}$ where base e was chosen.

Exercises for Section 5.2.

(1) When $a \leq 0$, why is $\log_b a$ undefined?

(2) Expand each expression as much as is possible.

(a) $\log_5(\sqrt{x} \cdot y^4)$

(b) $\ln \dfrac{x^4 \sqrt[3]{y}}{\sqrt{z}}$

(c) $\log_7 \dfrac{4}{xy}$

(d) $\log \dfrac{\sqrt{x} \cdot y}{z^3}$

(e) $\log_b \sqrt[4]{\dfrac{x^4 y^3}{z^5}}$

(f) $\log_a \dfrac{27a^6}{16y^2}$

(3) Write each expression as a single logarithm.

(a) $\dfrac{1}{2}\log_2 x - 3\log_2 y - 4\log_2 z$

(b) $3\log x - \dfrac{4}{3}\log y - 5\log z$

(c) $\log_3(x^2 - 16) - 2\log_3(x+4)$

(d) $\log_4(x^2 - x - 6) - \log_4(x^2 - 9)$

(4) Convert the following logarithmic equations into exponential equations.

(a) $\log_2 8 = 3$

(b) $\log_{\frac{1}{3}} 3 = -1$

(c) $\ln e = 1$

(d) $\log_a r = M$

(e) $\log_7 1 = 0$

(f) $\log 0.01 = -2$

(5) Convert the following exponential equations into logarithmic equations.

(a) $3^3 = 27$

(b) $e^0 = 1$

(c) $\sqrt{36} = 6$

(d) $\sqrt[3]{125} = 5$

(e) $\left(\dfrac{2}{3}\right)^{-1} = \dfrac{3}{2}$

(f) $B^r = P$

(6) Use the definition of a logarithm to solve the following equations.

(a) $\log_x 16 = -4$

(b) $\log_{125} x = \dfrac{2}{3}$

(c) $\log_4 (x^3) = 7$

(d) $x = e^{3 \ln 8}$

(7) Use the definition of a logarithm to determine the equation of the inverse function for each of the following. State the domain and range of the original function, and the domain and range of the inverse.

(a) $f(x) = e^{3x}$

(b) $g(x) = \ln\left(\dfrac{1}{x}\right)$

(c) $h(x) = \log_2 (x + 5)$

(d) $j(x) = 3^{x-4}$

(8) For each part of the previous problem, evaluate the compositions

$$f(f^{-1}(x)) \quad \text{and} \quad f^{-1}(f(x))$$

and state the domain on which each composition is the identity function. Use graphing software to verify your results.

(9) The pH scale is used to classify how acidic or basic a liquid solution may be. Mathematically, the scale is given by

$$pH = -\log H^+$$

where H^+ represents hydrogen ions in solution.

(a) Solve the pH relationship for H^+.

(b) Higher concentrations of hydrogen ions result in a more acidic solution, while lower concentrations of hydrogen ions result in a more basic solution. Do higher pH values mean the solution is more basic or acidic? Explain your reasoning.

(c) Suppose you have two solutions. One has a pH of 4, the other a pH of 6. Quantify the difference in the acidity between these two solutions.

5.3. Graphing Logarithmic Functions

The graph of the fundamental function $f(x) = \ln x$ is shown below.

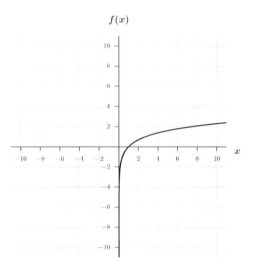

FIGURE 4. $f(x) = \ln x$

Some features worth noting are the domain, $\{x \mid x \in \mathbb{R} \text{ and } x > 0\}$, the range, $\{f(x) \mid f(x) \in \mathbb{R}\}$, and the vertical asymptote at $x = 0$. As the natural logarithm is the inverse of an exponential with base e, the functions $f(x) = \ln x$ and $g(x) = e^x$ would be symmetric across the identity function (graph them all together and verify). Further still, the domain and range of f would, respectively, be the range and domain of g (verify this as well).

All previous sections involving graphing transformations and function compositions are applicable to logarithms. As an example, let the function g be given by

$$g(x) = \ln(x + 3)$$

Our experiences in previous chapters tell us that this will be a transformation of the fundamental function $f(x) = \ln x$. Specifically, g is the function f shifted 3 units to the left. Figure 5 shows the graph of this transformation.

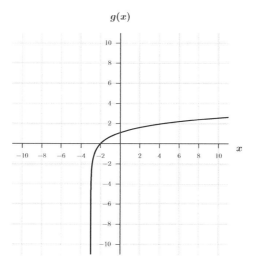

FIGURE 5. $g(x) = \ln(x+3)$

As another example, let us graph the function

$$f(x) = \log_2(x^2)$$

A reasonable first step would be to consider the domain of f. Note that we will square all inputs before the logarithm is taken. This means that the domain of f is $\{x \mid x \in \mathbb{R} \text{ and } x \neq 0\}$. Let us explore how f would act on a negative input value.

$$f(-x) = \log_2\left((-x)^2\right) = \log_2(x^2) = f(x)$$

Recall that the relationship $f(-x) = f(x)$ is the definition of even symmetry. This means that f is symmetric across the vertical axis. We need only to graph f for $x > 0$ and then reflect the graph across the vertical axis.

Three convenient input values to use for a logarithmic function with base b are $\frac{1}{b}$, 1, and b. For the function f this would give us

x	$f(x)$
$\frac{1}{2}$	$\log_2\left(\frac{1}{2}\right)^2 = -2$
1	$\log_2(1)^2 = 0$
2	$\log_2(2)^2 = 2$

from which we get the points $(\frac{1}{2}, -2)$, $(1, 0)$, and $(2, 2)$. The reflection of these points across the vertical axis would be $(-\frac{1}{2}, -2)$, $(-1, 0)$, and $(-2, 2)$. These six points, the asymptote at $x = 0$, and general knowledge on the overall shape of a logarithmic function are more than enough to get a good sketch of f.

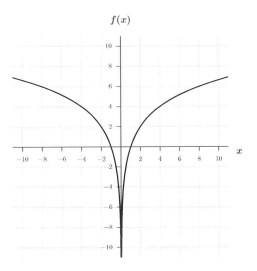

FIGURE 6. $f(x) = \log_2(x^2)$

The final example of graphing logarithmic functions will be

$$f(x) = \ln(x^2 + 3x - 4)$$

We can decompose f into two functions g and h where

$$g(x) = \ln x \qquad \text{and} \qquad h(x) = x^2 + 3x - 4$$

The composition of f would then be $f(x) = g(h(x))$. The order of this composition indicates that we are using the vertical output values of h as input to g. As we know that logarithmic functions must have inputs greater than zero, we begin by considering the zeros of h, $x = 1$ and $x = -4$.

As illustrated in Figure 7, $h(x) > 0$ for $x > 1$. The function

$$f(x) = g(h(x)) = \ln(h(x))$$

must be defined for $x > 1$ as we know that we can take the natural logarithm of positive real numbers. As x approaches 1 from the right, $h(x)$ approaches 0. Our previous experience with the natural logarithm function tells us that there must

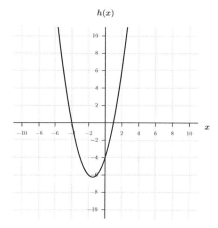

FIGURE 7. $h(x) = x^2 + 3x - 4$

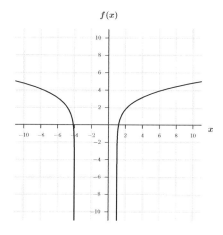

FIGURE 8. $f(x) = \ln(h(x))$

then be a vertical asymptote at $x = 1$. The exact same arguments tell us that $f(x) = g(h(x)) = \ln(h(x))$ must be defined for $x < -4$, and that there must be a vertical asymptote at $x = -4$. Additionally, the composition does not exist for domain values of $-4 \leq x \leq 1$. This is because $h(x) \leq 0$ for $-4 \leq x \leq 1$.

At this point, we have a general idea of the overall shape and behavior of $f(x) = g(h(x)) = \ln(x^2 + 3x - 4)$. We know that the domain of f is $x < -4$ or $x > 1$; we know that there are vertical asymptotes at $x = -4$ and $x = 1$; we know that for values of $x > 1$ and $x < -4$ the function is simply taking the natural logarithm of increasing positive numbers. This information is enough for us to draw a rough sketch of f, shown in Figure 8.

Note that we have not located any particular points on the graph of f. Doing so would require us to pick a value for $f(x)$ and solve the resulting equation for x. To illustrate this idea, suppose we wanted to locate the x-intercepts on the graph of f. We would need to solve the equation

$$0 = \ln(x^2 + 3x - 4)$$

This process will be the topic of the next section.

Exercises for Section 5.3.

(1) Each of the following functions is either some number of transformations on a general logarithmic function, or a composition of functions.

(a) $f(x) = -\ln(x+1)$

(d) $j(x) = 2\log(-x) + 7$

(b) $g(x) = \ln\left(\dfrac{1}{x}\right)$

(e) $k(x) = \ln\left(\dfrac{1}{|x|}\right)$

(c) $h(x) = -\dfrac{1}{4}\ln(x-3) + 3$

(f) $l(x) = \log(x^4 + 1)$

If the given function can be written as transformations on a general logarithmic function:

(a) State the general logarithmic function and the corresponding points of $\left(\tfrac{1}{b}, -1\right)$ and $(b, 1)$.

(b) List the transformations that result in the given function and perform the required transformations on the points $\left(\tfrac{1}{b}, -1\right)$ and $(b, 1)$.

(c) Verify that your transformed points satisfy the given function and use these points to sketch the transformed function.

If the given function is a composition of functions:

(a) Decompose the composition into two simpler functions.

(b) Sketch the graph of the inner function and use this graph to determine the domain of the composition.

(c) Roughly sketch the graph of the full composition. You do not have to list specific points, just the overall behavior (shape).

(d) If the graph of the composition appears to have even or odd symmetry, prove this result.

(2) Use graphing software to examine the graph of
$$f(x) = \log\left(\sqrt{x^2 + 1} + x\right)$$
Prove the symmetry relationship illustrated in the graph of f.

5.4. Solving Equations With Logarithms

Logarithms can be used to solve equations where the variable we wish to solve for is an exponent (or part of an exponent). All of the previously established rules and properties of logarithms are at our disposal, we only need apply them. Suppose we wanted to solve the equation

$$10^{3x} = 2.1$$

There are two equivalent ways to approach this type of equation. The first would be to use the definition of a logarithm to convert this exponential equation into logarithmic form.

$$\log(2.1) = 3x$$

Make sure you understand the relationship between the exponential equation we started with and this logarithmic version. As 10 to the power of $3x$ is 2.1, it must be true that the logarithm base 10 of 2.1 is $3x$. Solving for x is now simply division.

$$x = \frac{\log(2.1)}{3} \approx 0.107$$

Though the decimal approximation of this answer is interesting, we will usually not reduce logarithms to decimal answers in the future. The exception will be applied problems in which a meaningful decimal answer is required.

The second way to approach an equation of that type would be to take the logarithm of both sides. Let us solve

$$6^x = 18$$

Before we take the logarithm, we need to decide which base to use. Should we use base 10, base 6, base e? Does it matter? The steps below illustrate the results of using base 10 and base e.

$$\log(6^x) = \log(18) \qquad\qquad \ln(6^x) = \ln(18)$$

$$x \log(6) = \log(18) \qquad\qquad x \ln(6) = \ln(18)$$

$$x = \frac{\log(18)}{\log(6)} \qquad\qquad x = \frac{\ln(18)}{\ln(6)}$$

5.4. SOLVING EQUATIONS WITH LOGARITHMS

If you run both of these answers through a calculator, you will see that the decimal approximations are exactly the same. In fact, we can use a change of base to verify these answers are identical.

$$\frac{\log(18)}{\log(6)} = \frac{\frac{\ln(18)}{\ln(10)}}{\frac{\ln(6)}{\ln(10)}} = \frac{\ln(18)}{\ln(10)} \cdot \frac{\ln(10)}{\ln(6)} = \frac{\ln(18)}{\ln(6)}$$

Remember, you are free to use any base you wish but base 10 and base e are extremely common and useful choices.

Another type of equation is when the variable is inside a logarithm.

$$\log(5x) = 3$$

Again, there are two equivalent ways to approach this problem. The first is to use the definition of the logarithm. If the logarithm base 10 of $5x$ is 3, then it must be true that

$$5x = 10^3$$

Solving for x will leave you with $x = 200$.

The second way to approach equations of this type is to treat both sides of the equation as if they are exponents on a base (this was another property discussed in an earlier section). Consider the following.

$$\log(x - 1) + \log(x - 4) = 1$$

First off, we can use the properties of the logarithm to rewrite this as

$$\log[(x - 1)(x - 4)] = 1$$

Now, we can treat both sides of this equation as though they are exponents on a base of our choosing. We choose base 10.

$$10^{\log[(x-1)(x-4)]} = 10^1$$

The reasoning behind the choice of base 10 was that now (using another property of logarithms) the equation simplifies to

$$(x - 4)(x - 1) = 10^1$$

Note that this is exactly the same exponential relationship that is described in the logarithmic equation $\log[(x-1)(x-4)] = 1$. We could have also arrived here using only the definition of the logarithm. Solving for x gives

$$(x-4)(x-1) = 10$$

$$x^2 - 5x + 4 = 10$$

$$x^2 - 5x - 6 = 0$$

$$(x-6)(x+1) = 0$$

and we arrive at $x = 6$ or $x = -1$. Both answers must be checked to make sure they satisfy the domain of the logarithm function. Checking $x = 6$ yields

$$\log(6-1) + \log(6-4) = 1$$

$$\log(5) + \log(2) = 1$$

As both the logarithm of 5 and 2 are defined, $x = 6$ is a valid answer. Checking $x = -1$ yields

$$\log(-1-1) + \log(-1-4) = 1$$

$$\log(-2) + \log(-5) = 1$$

In this case, one or more logarithms is undefined (recall that we can only take the logarithm of values greater than zero). This means that $x = -1$ is not a solution to the equation.

Our next example is similar to the previous and solved below.

$$\ln(x^2 - x - 6) - \ln(x+2) = 2$$

$$\ln\left(\frac{x^2 - x - 6}{x+2}\right) = 2$$

$$\ln\left(\frac{(x+2)(x-3)}{x+2}\right) = 2$$

$$\ln(x-3) = 2$$

$$x - 3 = e^2$$

$$x = e^2 + 3$$

5.4. SOLVING EQUATIONS WITH LOGARITHMS

We leave it to the reader to verify that the solution satisfies the domain of the natural logarithm.

Let us return to a problem that was posed several sections ago. Given that the half-life of carbon-14 is $5,730$ years, we have previously shown that given a starting amount of carbon-14, A_0, this decay can be modeled by

$$A(N) = A_0 \left(\frac{1}{2}\right)^{\frac{N}{5730}}$$

where N is years. As an example, if you had 25kg of carbon-14 and let $15,700$ years pass, you would be left with

$$A(15,700) = (25)\left(\frac{1}{2}\right)^{\frac{15,700}{5,730}} \approx 3.742 \text{ kg}$$

Our recent developments with logarithms allow us to answer additional questions involving these types of models. Suppose we found a human bone fragment that contained approximately 23% of the carbon-14 normally found in a living person. We can now determine the age of the bone.

The steps below illustrate solving for the amount of time required for this amount of decay. Note that we do not know exactly how much carbon-14 we started with. What we do know is that the amount we found is approximately 23% the amount that would be in a living person, or $(0.23)A_0$.

$$0.23(A_0) = (A_0)\left(\frac{1}{2}\right)^{\frac{N}{5,730}}$$

$$0.23 = \left(\frac{1}{2}\right)^{\frac{N}{5,730}}$$

$$\ln(0.23) = \ln\left(\left(\frac{1}{2}\right)^{\frac{N}{5,730}}\right)$$

$$\ln(0.23) = \left(\frac{N}{5,730}\right)\ln\left(\frac{1}{2}\right)$$

$$\frac{\ln(0.23) \cdot 5,730}{\ln\left(\frac{1}{2}\right)} = N$$

$$12,149.285 \approx N$$

Thus, our mysterious bone fragment is approximately $12,149$ years old.

As a final thought in the exponential modeling of half-life, let us return to our model for carbon-14 decay and explore a bit. Our previous work with logarithms provides us with the relationship

$$e^{\ln \frac{1}{2}} = \frac{1}{2}$$

Suppose we make this substitution for the $\frac{1}{2}$ in the carbon-14 model.

$$A(N) = A_0 \left(\frac{1}{2}\right)^{\frac{N}{5,730}} = A_0 \left(e^{\ln \frac{1}{2}}\right)^{\frac{N}{5,730}} = A_0 e^{\left(\frac{\ln \frac{1}{2}}{5,730}\right)N}$$

This substitution (and the resulting algebra) offer quite an interesting thought. Namely, this decay can be accurately modeled by either of the two exponential models we previously developed.

Exercises for Section 5.4.

(1) Solve the following equations.

(a) $4^{x+3} = e^x$

(b) $\log_x 14 = 7$

(c) $e^{x^2} = 13$

(d) $7^{2x-3} = 9^{x-1}$

(e) $\log_2(x+1) + \log_2(x+3) = \log_2 5$

(f) $\ln(x^4) = (\ln x)^3$

(g) $3^{x^2-1} = 5$

(h) $4^{2x} - 3(4^x) - 10 = 0$

(2) Solve for the specified variable.

(a) k: $\quad Be^k t - R = Q$

(b) W: $\quad \log_n \left(\dfrac{W}{L}\right) = P$

(c) Z: $\quad \log_W (B + e^2) = \log_W \left(\dfrac{J}{Z}\right)$

(d) H : $\log_H (D^4 + F) = \dfrac{\log R}{S}$

(e) r : $Pe = A \ln (Dr + B)$

(f) a : $U = \dfrac{\ln b}{Ne^{-Pa} + m}$

(3) Determine the equation of the inverse function for each of the following. State the domain and range of the original function and the domain and range of the inverse.

(a) $f(x) = 2 \cdot \ln(x+1)$

(b) $g(x) = 3e^{5x}$

(c) $h(x) = \dfrac{1}{4} 10^{x^2}$

(d) $j(x) = \ln(x^2 - 7)$

(4) The half-life of titanium-44 is sixty three years. Using a base of one half, develop a model for the decay of titanium-44. Using base e, develop a model for the decay of titanium-44. Use exact values; do not round. Once both models are complete, prove that they are in fact the same model with different bases.

Now, suppose you have some titanium-44. How much of your starting amount will remain after 15 years? You may answer with either model, as you have already proven them to be the same. Comment on your choice of model.

CHAPTER 6

Polynomial, Rational, and Power Functions

6.1. Polynomial Functions

Before we can discuss polynomial functions, we must first formalize the definition of a polynomial. A *polynomial in a single variable* is a term, or the sum of a finite number of terms, of the form

$$ax^n$$

where $a \in \mathbb{R}$, n is an integer and $n \geq 0$. The *degree* of a polynomial in one variable is the largest exponent (value of n) in the polynomial. Some simple examples of polynomials are

$$x^2 \qquad 3x^7 + 9x^3 + 1 \qquad \frac{2}{5}x^{12} + 4$$

Note that we are only considering polynomials in a single variable, x. Multivariate polynomials do exist, but we will not discuss them here.

Two of our fundamental functions are polynomials

$$f(x) = x^2 \quad \text{and} \quad g(x) = x^3$$

These two specific examples will be used to help generalize *end behavior*. The end behavior of a polynomial function is the behavior of the output as x increases without bond, $x \to \infty$, and as x decreases without bound, $x \to -\infty$.

In the case of $f(x) = x^2$, we know that as $x \to \infty$, the output, $f(x)$, also increases without bound, $f(x) \to \infty$. Similarly, as $x \to -\infty$, the output still increases without bound, $f(x) \to \infty$. Spend a moment with the graph of f and verify these relationship. Similarly, for $g(x) = x^3$, the end behavior would be $g(x) \to \infty$ as $x \to \infty$, and $g(x) \to -\infty$ as $x \to -\infty$. Verify these relationships on the graph of g.

Generally speaking, the end behavior of an nth degree polynomial depends on n being even, or odd, and on the coefficient of the nth degree term. If n is even, and the coefficient of the nth degree term is positive, the polynomial will have the same end behavior as $f(x) = x^2$. If n is even, and the coefficient of the nth degree term is negative, then the polynomial will have the same end behavior of $-f(x) = -x^2$.

Consider the polynomials

$$h(x) = 3x^4 + x^3 - 2x + 1 \quad \text{and} \quad j(x) = -2x^6 + x^3 + 4$$

The graphs of h and j are shown in Figure 1. The function h exhibits the same end behavior of $f(x) = x^2$, while the function j matches the end behavior of $-f(x) = -x^2$.

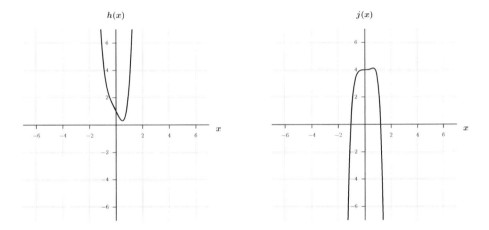

FIGURE 1. The functions $h(x) = 3x^4 + x^3 - 2x + 1$ and $j(x) = -2x^6 + x^3 + 4$

Similarly, if n is odd and the coefficient of the nth degree term is positive, the polynomial will exhibit end behavior similar to $g(x) = x^3$. If n is odd and the coefficient of the nth degree term is negative, the polynomial will exhibit end behavior similar to $-g(x) = -x^3$.

Consider the polynomials

$$h(x) = x^5 + 2x^3 - 8x^2 + 3 \quad \text{and} \quad j(x) = -x^5 + 2x^4 - x^2 + 1$$

The graphs of h and j are shown in Figure 2. The function h exhibits the same end behavior of g, while the function j matches the end behavior of $-g(x) = -x^3$.

A topic of particular interest will be the *zeroes* of a polynomial. Recall that the zeroes are the values of x that result in $f(x) = 0$. Graphically, the zeroes of a polynomial are where the graph intersects the x-axis.

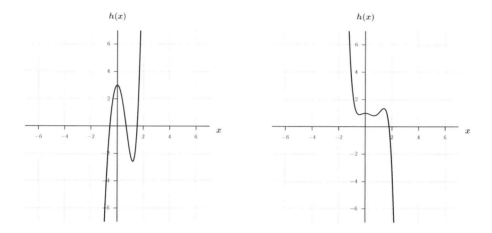

FIGURE 2. The functions $h(x) = x^5 + 2x^3 - 8x^2 + 3$ and $j(x) = -x^5 + 2x^4 - x^2 + 1$

Consider the polynomial function

$$f(x) = x^2 - 5x + 6$$

Solving for the zeroes of this parabola is fairly trivial, but we will go through the steps as they will help us to develop a new idea.

$$0 = x^2 - 5x + 6$$
$$= (x-3)(x-2)$$

The zero factor property allows us to conclude that either

$$x - 3 = 0 \quad \text{or} \quad x - 2 = 0$$

and so we have that

$$x = 3 \quad \text{or} \quad x = 2$$

These steps, in conclusion with the final answers for x illustrate the foundation of a very powerful theorem for polynomials, the *Factor Theorem*.

THEOREM. *Let f be a polynomial function. Then f has a factor of $(x-a)$ if and only if $f(a) = 0$.*

The Factor Theorem may seem like overkill for identifying the zeros of a parabola, and in truth, it is. Solving for the zeroes of any second degree polynomial is not difficult, but what about third, fourth, fifth, and n^{th} degree polynomials? The Factor Theorem works on all polynomials, regardless of the degree.

Suppose we wish to factor and identify the zeroes of the polynomial function

$$g(x) = x^3 - 5x^2 + 5x - 1$$

The Factor Theorem guarantees that if we can identify a value of x such that $g(x) = 0$, we will have also identified a factor of g. Note that $g(1) = 0$. This means that $x - 1$ must be a factor of g. We can use this fact to help us identify the other zeroes of g through polynomial long division.

$$
\begin{array}{r}
x^2 - 4x + 1 \\
x - 1 \overline{\smash{)}\ x^3 - 5x^2 + 5x - 1} \\
\underline{-x^3 + x^2 } \\
-4x^2 + 5x \\
\underline{4x^2 - 4x } \\
x - 1 \\
\underline{-x + 1} \\
0
\end{array}
$$

This division has shown

$$g(x) = x^3 - 7x^2 + 13x - 3 = (x - 1)(x^2 - 4x + 1)$$

The remaining zeroes of g are where $x^2 - 4x + 1 = 0$. We can use the quadratic formula to solve for these values of x.

$$x = \frac{-(-4) \pm \sqrt{(-4)^2 - 4(1)(1)}}{2(1)}$$

$$= \frac{4 \pm \sqrt{12}}{2}$$

$$= \frac{4 \pm 2\sqrt{3}}{2}$$

$$= 2 \pm \sqrt{3}$$

from which we can conclude

$$g(x) = x^3 - 7x^2 + 13x - 3 = (x - 1)\left(x - (2 + \sqrt{3})\right)\left(x - (2 - \sqrt{3})\right)$$

and we know that the zeroes of g are

$$x = 1, \quad x = 2 + \sqrt{3}, \quad x = 2 - \sqrt{3}$$

While extremely helpful, the Factor Theorem is not an ideal method of factoring a polynomial function as factors are only identified when zeroes are identified, and zeroes are generally identified through a process of trial and error. In the previous example the first zero was provided seemingly out of thin air to help illustrate the overall process. In reality, there is no known efficient method for factoring (or finding the zeroes) of an arbitrary n^{th} degree polynomial.

Once the zeros of a polynomial are identified, we can determine if the graph crosses, or simply intersects the x-axis at each zero. Consider the polynomial

$$f(x) = (x+2)^2 (x-1)$$

As f is factored, it is clear that the zeroes are

$$x = -2 \quad \text{or} \quad x = 1$$

A sign chart will illustrate weather or not the graph crosses at these respective zeroes.

		-2		1	
$(x+2)^2$	$+$	0	$+$	$+$	$+$
$x-1$	$-$	$-$	$-$	0	$+$
$(x+2)^2(x-1)$	$-$	0	$-$	0	$+$

The last row of the sign chart tells us that the output of f is negative to the left of $x = -2$, and negative to the right of $x = -2$. Thus, the graph of f intersects, but does not cross the x-axis at $x = -2$. Conversely, the output of f is negative to the left of $x = 1$, and positive to the right of $x = 1$. This means that the graph of f crosses the x-axis at $x = 1$. The graph of f is shown in Figure 3.

The behavior of f around its zeroes is a direct result of the exponent associated with each factor of f. Any factor being raised to an even exponent is said to have *even multiplicity*. The graph of a function cannot cross a zero that is the result of a factor with even multiplicity. Similarly, factors being raised to odd exponents are said to have *odd multiplicity*. The graph of a function will cross zeros that are the result of a factor with odd multiplicity.

We can also select specific zeroes we would like a polynomial function to have, and determine an equation of a polynomial with the selected zeroes. Suppose we are

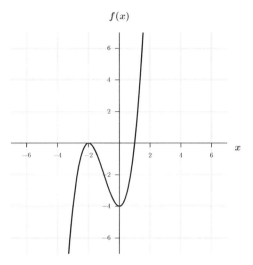

FIGURE 3. The function $f(x) = (x+2)^2 (x-1)$

interested in the specific zeroes of $x = 3$, $x = 11$, and $x = -5$. The Factor Theorem tells us that a polynomial function, f, with those specific zeroes will have the form

$$f(x) = (x-3)(x-11)(x+5)$$

While this is the simplest possible choice for f, it is not the only choice. This function could be vertically stretched or compressed without changing its zeroes. So f could be also be written

$$f(x) = a(x-3)(x-11)(x+5)$$

where $a \in \mathbb{R}$. Further still, we did not specify the multiplicity of the zeroes, so f could have the form

$$f(x) = a(x-3)^l (x-11)^m (x+5)^n$$

where l, m, and n are positive integers. In summary, there are infinite possible choices for the function f.

Suppose we wish to write the equation of a polynomial function of least degree such that it contains the points

$$(-3, 0),\ (2, 0),\ (4, 0),\ (1, 5)$$

The polynomial must be of the form

$$g(x) = a(x+3)(x-2)(x-4)$$

While we could put various exponents on all of these factors, doing so would increase the degree of the polynomial. As we want a polynomial of least degree, it must be the case that all of the exponents on the non-constant factors are equal to one. The current equation for the pairings of g will intersect all of the required zeroes. We can use the remaining point $(1,5)$ to solve for a.

$$g(1) = a(1+3)(1-2)(1-4)$$

$$5 = a(4)(-1)(-3)$$

$$5 = 12a$$

$$\frac{5}{12} = a$$

So the final equation for the pairings of g will be

$$g(x) = \frac{5}{12}(x+3)(x-2)(x-4)$$

Exercises for Section 6.1.

(1) Describe the end behavior of the following polynomial functions.

(a) $f(x) = 2x^7 - 4x^5 + 3x^3 - 9$

(b) $g(x) = -x^6 + x^5 - 13x^2$

(c) $h(x) = -x^{13} - 20x^9 + 13x^5 + 2$

(d) $j(x) = 19x^6 + 7x^5 - 10$

(2) Identify the zeros of the following polynomial functions, and determine if the function changes sign at each zero. You may need to utilize the Factor Theorem to identify an initial zero.

(a) $f(x) = x(x-3)^2(x+2)^3(x+11)^4$

(b) $g(x) = (x+4)(x^2+5x+6)$

(c) $h(x) = (x-7)(x^2+3x+1)$

(d) $j(x) = x^3 + x^2 - 8x - 12$

(3) Determine the polynomial function of least degree, such that it passes through the given points.

(a) $(2,0)$, $(4,0)$, $(-3,0)$

(b) $(-4,0)$, $(0,0)$, $(2,3)$

(c) $(-1,0)$, $(1,0)$, $(2,0)$, $(3,4)$

(d) $(-2,0)$, $(0,0)$, $(a,0)$, $(3,3)$

(4) Determine the equation of a cubic polynomial function such that it has the following zeroes

$$x = 5 \qquad x = 2 \pm \sqrt{3}$$

Your final answer should be in the form $f(x) = ax^3 + bx^2 + cx + d$.

(5) Our previous work in linear modeling illustrated that the linear regression line minimizes the sum of the squared distance from all given data values. Suppose instead, we want a function that passes through all given data values. The function would need to be non-linear, as there is no plausible way that the given data would be points on a line. Let our data consist of the four points

$$(-1, -2), (0, 7), (1, 4), (2, 3)$$

How might we find a function that passes through these four points? Consider a generic polynomial function of degree three.

$$f(x) = ax^3 + bx^2 + cx + d$$

This equation has four unknowns, a, b, c, d. We can utilize the four known points to solve for these four unknown values. The resulting polynomial will pass through all four of the given points. To do so, we substitute the four points into the generic third degree polynomial. The result is a system of four equations in four variables.

$$-a + 4b - c + d = -2$$
$$0a + 0b + 0c + d = 7$$
$$a + b + c + d = 4$$

$$8a + 4b + 2c + d = 3$$

Solve this system for a, b, c, d, and write the equation of the polynomial function. Use graphing software to verify that the polynomial passes through the four given points.

(6) Write a possible equation for each of the graphs given below. You may leave your answer in factored form and assume plotted points have integer value coordinates.

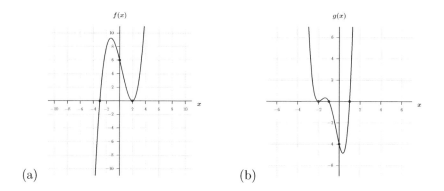

(a)　　(b)

6.2. Rational Functions

A *rational function* is a function that is written as the ratio of two polynomials. Put simply, a function f is rational if the pairings of f are given by an equation of the form

$$f(x) = \frac{\text{some polynomial}}{\text{some other polynomial}}$$

The purpose of this section is to explore the general behavior of rational functions. Specifically, we will consider five topics:

(1) Zeroes

(2) Vertical asymptotes

(3) Horizontal asymptotes

(4) Oblique asymptotes

(5) Holes

Zeroes and vertical asymptotes are ideas that have previously been introduced. Given a rational function it seems reasonable to assume that the zeroes would be input values that make the numerator equal to zero, and vertical asymptotes would occur at the input values that make the denominator equal to zero. These ideas are correct with one exception, the special case of *holes*, which will be discussed shortly.

Horizontal and *oblique* asymptotes describe the end behavior of rational functions. That is to say that these types of asymptotes quantify what happens to the output as $x \to -\infty$ and as $x \to \infty$. The end behavior of a rational function is determined by the degree of the numerator and the degree of the denominator.

Suppose we have a function f such that

$$f(x) = \frac{ax^m + \ldots}{bx^n + \ldots}$$

- If $m < n$ then f has a horizontal asymptote at a vertical value of 0.

- If $m = n$ then there is a horizontal asymptote at a vertical value of $\frac{a}{b}$.

- If $m = n + 1$ then there is an oblique asymptote that is found through polynomial long division.

The function f has at most one horizontal or oblique asymptote. It may not have both. Let us consider an example of each case, starting with the fundamental function

$$f(x) = \frac{1}{x}$$

In this example, the degree of the numerator is equal to zero, while the degree of the denominator is equal to one. This tells us that we expect a horizontal asymptote, $y = 0$, and our previous experience with this fundamental function verifies this expectation.

Next, we have a function g such that

$$g(x) = \frac{3x^2 + 1}{3x^2 - 2x + 1}$$

The degree of the numerator and denominator are both equal to two, so we expect a horizontal asymptote at $y = \frac{a}{b}$ where a is the coefficient of the term with the highest degree in the numerator, and b is the coefficient of the term with the highest degree in the denominator. In this example, both a and b are equal to three, so the horizontal asymptote will be $y = \frac{3}{3} = 1$. The graph of g is shown in Figure 4, with the horizontal asymptote shown as a dotted line. Note that the graph of g crosses the horizontal asymptote. Unlike vertical asymptotes, the graph of a function may intersect or cross a horizontal or oblique asymptote.

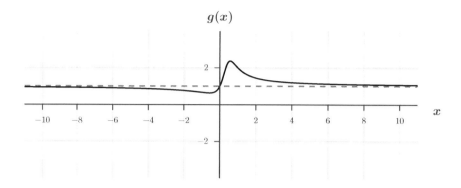

FIGURE 4. The function $g(x) = \frac{3x^2+1}{3x^2-2x+1}$. Horizontal asymptote $y = 1$ shown by dotted line.

Finally, consider the function h such that

$$h(x) = \frac{x^2 - x - 6}{x - 2}$$

In this example, the numerator has degree two, while the denominator has degree one. This tells us to expect an oblique asymptote. The equation of the oblique asymptote is found through polynomial long division.

$$\begin{array}{r}
x + 1 \\
x - 2 \overline{\smash{)}\, x^2 - x - 6 } \\
-x^2 + 2x \\
\hline
x - 6 \\
-x + 2 \\
\hline
-4
\end{array}$$

The remainder plays no part in the equation of the oblique asymptote, and can be ignored. Thus, the oblique asymptote has equation $y = x + 1$. Figure 5 illustrates the function h, and its oblique asymptote. Though not shown with a dotted line, the function h also has a vertical asymptote at $x = 2$.

Finally, we arrive at the topic of holes. A rational function will have a hole if the numerator and denominator both have the same polynomial factor. As an example, consider the function

$$f(x) = \frac{(x+2)(x-1)}{x+2}$$

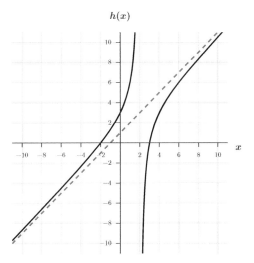

FIGURE 5. The function $h(x) = \frac{x^2-x-6}{x-2}$. Oblique asymptote $y = x+1$ shown by dotted line.

As the domain of this function is $x \neq 2$, the factor of $x+2$ in the numerator and denominator is not equal to zero, and we are allowed to simplify the fraction.

$$f(x) = \frac{\cancel{(x+2)}(x-1)}{\cancel{x+2}} = x - 1$$

At this step we must recall a previously mentioned, and extremely important fact: a function is defined by exactly what is given, not by how it may be simplified. As defined, the function f cannot have a domain value of $x = -2$, as that would result in division by zero. The unallowed domain value of $x = -2$ corresponds to a hole in the function f. Visually, f is the straight line given by $f(x) = x - 1$, but there is a literal hole at $x = -2$ as the domain of f is

$$\{x \mid x \in \mathbb{R} \text{ and } x \neq -2\}$$

Figure 6 illustrates the graph of f. A function may have multiple holes with any combination of vertical or horizontal asymptotes.

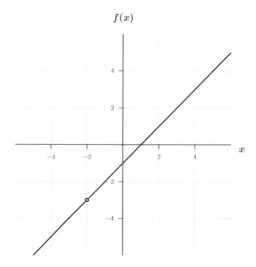

FIGURE 6. The function $f(x) = \frac{(x+2)(x-1)}{x+2}$ with hole shown at $x = -2$.

Exercises for Section 6.2.

(1) Describe the end behavior of the following rational functions. Use graphing software to verify your results.

(a) $f(x) = \dfrac{9x^3 + 2x^2 - 11}{4x^3 + 3x + 1}$

(b) $g(x) = \dfrac{x^{17} + 3}{x^{19} + 3}$

(c) $h(x) = \dfrac{6x^5 + 8x^3 - 7}{4x^2 - 3x + 4}$

(d) $j(x) = \dfrac{2x^3 + 3x^2 + 8x - 5}{x^2 - 3}$

(2) Describe end behavior, identify all intercepts, asymptotes, and holes of the following rational functions. You may need to utilize the Factor Theorem.

(a) $f(x) = \dfrac{x^2 + 4x + 3}{x^2 - x - 2}$

(b) $g(x) = \dfrac{2x^3 - 6x^2 - 8x}{x^2 - 2x}$

(c) $h(x) = \dfrac{3x^3 - 12x^2 - 36x}{x^3 + 3x^2 - 4x - 12}$

(d) $j(x) = \dfrac{x^2 - 2}{-2x - 4}$

(3) Each of the following functions intersects its non-vertical asymptote. Solve for the coordinates of the intersection.

(a) $f(x) = \dfrac{x^2 - 3x + 1}{x^2 + 9x}$ \qquad (b) $g(x) = \dfrac{x^3}{x^2 + 3x - 5}$

6.3. Power Functions

The general form of a *power function* is given by

$$f(x) = kx^p$$

where $p, k \in \mathbb{R}$. The constant k is referred to as the *constant of proportionality*, while the constant p is the exponent on the input x. The sign of p determines the characterization of proportionality.

- If $p > 0$, then $f(x)$ is *proportional* or *directly proportional* to the pth power of x.

- If $p < 0$, then $f(x)$ is *inversely proportional* to the pth power of x.

Some examples illustrating these characterizations are

$$f(x) = kx^5 \quad \text{and} \quad g(x) = kx^{-2} = \dfrac{k}{x^2}$$

In these examples, $f(x)$ is proportional to the 5th power of x, and $g(x)$ is inversely proportional to the square of x.

Several of our fundamental functions are power functions.

$$f(x) = x^{-1} = \dfrac{1}{x} \qquad g(x) = x^2 \qquad h(x) = x^3 \qquad j(x) = x^{\frac{1}{2}} = \sqrt{x}$$

Note that functions g and h are also polynomial functions. Polynomial functions are the sum of power functions where the exponent, p, is a natural number or zero.

Suppose we have a power function, f, that passes through the points $(32, 16)$ and $(1, 2)$. We can use this information to determine the equation for $f(x)$. Starting with the general form of a power function

$$f(x) = kx^p$$

we know that the two given points will result in the equations

$$16 = k \cdot 32^p \quad \text{and} \quad 2 = k \cdot 1^p$$

The equation on the right tells us that $k = 2$, and we can use this information in the equation on the left.

$$16 = k \cdot 32^p$$

$$16 = 2 \cdot 32^p$$

$$8 = 32^p$$

$$\ln(8) = p \cdot \ln(32)$$

$$\frac{\ln 8}{\ln 32} = p$$

$$0.6 = p$$

$$\frac{3}{5} = p$$

At this point, we know all constant values in the power function. The final version of the function is given by

$$f(x) = 2x^{\frac{3}{5}}$$

Power and exponential functions may seem similar, but they are extremely different. Be careful not to confuse them.

Exponential Functions	Power Functions
$g(x) = ab^x$	$f(x) = kx^p$
Input becomes an exponent on the base	Input becomes the base

To illustrate the difference in how exponential and power functions increase, consider the functions f and g given by

$$f(x) = 2^x \quad \text{and} \quad g(x) = x^3$$

The left half of Figure 7 suggests that the output of $f(x) = 2^x$ starts off above the output of $g(x) = x^3$, but then $g(x)$ overtakes and remains above $f(x)$ from then on. In fact, this is incorrect. As illustrated in in the right half of Figure 7, the exponential output ultimately catches and overtakes the power function output. Though we will not prove it here, any exponential function with base greater than 1 will ultimately overtake any power function regardless of the exponent the power function utilizes.

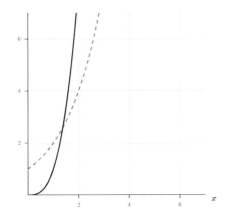 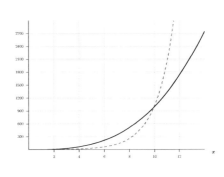

FIGURE 7. The functions $f(x) = 2^x$ (dashed) and $g(x) = x^3$ (solid).

Solving for the location of this intersection would mean setting the mathematics responsible for the pairings of f equal to the mathematics responsible for the pairings of g and solving for x. In other words

$$2^x = x^3$$

Spend a few minutes trying to solve this relationship for x. Though it is beyond the scope of this text, this equation cannot be algebraically solved for x.

Exercises for Section 6.3.

(1) For each of the following functions, state the domain and classify the function as exponential, polynomial, rational, or power. The variable a is a natural number, while b and c are both positive real numbers. Some functions may have multiple classifications.

(a) $f(x) = \dfrac{c}{ab}x + x$

(b) $g(x) = \dfrac{c+d}{x} - a$

(c) $h(x) = \dfrac{(abx)^{-1}}{c} - \dfrac{b^2}{x}$

(d) $j(x) = ab^x$

(e) $k(x) = ax^{-\frac{1}{2}}$

(f) $l(x) = \dfrac{a}{x^3} + \dfrac{bx^4}{a}$

(2) The functions f and g are defined below. As $x \to \infty$, which function's output will be larger?

$$f(x) = e^x \qquad g(x) = 400x^3$$

Use graphing software to verify your guess. Describe the behavior of these two functions and give an approximation of where they intersect.

(3) The gravitational force between two objects is inversely proportional to the square of the distance between their centers of mass. The constant of proportionality is a product of both object masses and the universal gravitational constant, $G = 6.673 \times 10^{-11} \; \frac{\text{N m}^2}{\text{kg}^2}$.

(a) Using distance as the input, develop a model for the gravitational force between two objects.

(b) Suppose all of space is empty, with the exception of two point mass objects. The first has a mass of 10^4 kg, and the second has a mass of 10^8 kg. They are separated by a distance of 2.3×10^8 meters. What is the gravitational force between them?

(c) Suppose once again that we have two point mass objects in empty space. Their masses are 1.3×10^4 kg and 4.5×10^5 kg, and the gravitational force between them is 2.498×10^{-5} N. How far apart are the point mass objects?

(d) Calculate the gravitational force between you and the center of the Earth.

(4) Determine the equation of the power function that passes through the given points.

(a) $(3, 1)$ and $(12, 5)$

(b) $(2, 3)$ and $(5, 9)$

CHAPTER 7

An Introduction to Trigonometric Functions

Our work thus far has provided a foundation for what comes next. If you find yourself unsure on any of the material in the preceding chapters, now is the time to go back and clear up any questions. Continuing forward without understanding how we arrived at this point will almost certainly result in unpleasantness.

Trigonometry is the study of triangles, and how their sides and angles relate to one another. We will use the relationships of triangles to extend our studies into circles.

7.0.1. Special Right Triangles. Consider the equilateral triangle $\triangle ABC$ shown below.

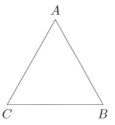

As $\triangle ABC$ is equilateral, we know that angles $\angle A$, $\angle B$, and $\angle C$ are all $60°$, and that all sides of the triangle have the same length. Let us draw a line from $\overline{CB}$ such that it bisects $\angle A$ (note that this line will also bisect $\overline{CB}$).

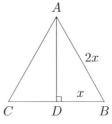

The result of this bisection is right triangle $\triangle ABD$, and the angles in $\triangle ABD$ are respectively $30°$, $60°$, and $90°$. If we refer to the length of $\overline{DB}$ as x, then it must be the case that the length of $\overline{AB}$ is $2x$. We can use the pythagorean theorem to solve for the length of $\overline{AD}$.

$$(\overline{AD})^2 + x^2 = (2x)^2$$

$$(\overline{AD})^2 = 4x^2 - x^2$$

$$\overline{AD} = \sqrt{3x^2}$$

$$\overline{AD} = x\sqrt{3}$$

We now know all side lengths and all angles of triangle $\triangle ABD$.

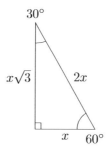

FIGURE 1. The 30-60-90 right triangle

This triangle is the first of the special right triangles and is commonly referred to as the *30-60-90 right triangle*.

The derivation of our second special right triangle begins with isosceles right triangle $\triangle ABC$ shown below.

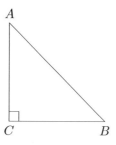

FIGURE 2. Isosceles right triangle $\triangle ABC$

As $\triangle ABC$ is isosceles, we know that $\overline{AC}$ and $\overline{CB}$ have the same length, and that $\angle A = \angle B = 45°$. If we refer to the length of $\overline{AC} = \overline{BC}$ as x, the pythagorean theorem will allow us to solve for the length of $\overline{AB}$.

$$x^2 + x^2 = (\overline{AB})^2$$

$$2x^2 = (\overline{AB})^2$$

$$\sqrt{2x^2} = \overline{AB}$$

$$x\sqrt{2} = \overline{AB}$$

All angles and lengths of triangle $\triangle ABC$ are shown below.

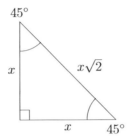

FIGURE 3. The 45-45-90 right triangle

This triangle is another of the special right triangles. It is commonly referred to as the *45-45-90 right triangle*. The 30-60-90 and 45-45-90 right triangles will play a large role in the upcoming sections.

7.1. Radians, Sine and Cosine

Degrees are a useful and meaningful unit of angle measurement (the previous section of this text utilized degrees in the derivation of the special right triangles), but they are somewhat inelegant when it comes to the study of trigonometry. While we could work through all of trigonometry in degrees, there is a much more convenient unit of measurement at our disposal.

Consider a circle of radius r that is centered at the origin. We will label the origin O, the x-intercept A, and draw a straight line from the origin into the first quadrant until it intersects the circle. The point where this line intersects the circle will be labeled B. This scenario is depicted in Figure 4.

The arc on the circle between points A and B is referred to simply as "arc AB". This arc corresponds to the angle $\angle AOB$. A ratio of the length of arc AB and the radius of the circle is the number of *radians* in angle $\angle AOB$. The variable theta, θ, is traditionally used to represent radians.

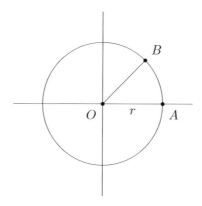

FIGURE 4

DEFINITION 7.1. Radians

For a circle of radius r, the number of radians associated with an arc AB is given by

$$\theta = \frac{\text{arc length of } AB}{r}$$

A full revolution around any circle would have arc length of $2\pi r$ (the circumference). The number of radians associated with a full revolution would be given by

$$\theta = \frac{2\pi r}{r} = 2\pi$$

In degrees, a full revolution is $360°$. The ratio of radians to degrees would be given by

$$\frac{2\pi}{360°} = \frac{\pi}{180°}$$

and we can use this ratio to convert measurements from degrees to radians. As an example, consider $90°$.

$$90° \left(\frac{\pi}{180°}\right) = \frac{\pi}{2}$$

To convert angles from radians to degrees, you would use the reciprocal. Specifically, $\frac{180°}{\pi}$. As an example, consider $\theta = \frac{3\pi}{4}$.

$$\left(\frac{3\pi}{4}\right)\left(\frac{180°}{\pi}\right) = 135°$$

Positive measurements of radians and angles always start from 0 and move in the counter-clockwise direction. Negative measurements of radians and angles always start from 0 and move in the clockwise direction. This idea is illustrated in Figure 5 below.

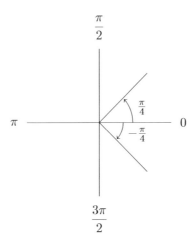

FIGURE 5. Positive and negative radians.

Dealing with angles larger than 2π is analogous to dealing with angles larger than $360°$, simply continue as far as is needed. A $540°$ measurement would result in a full revolution, then an additional half revolution. Similarly, a measurement of 3π would result in one full revolution and then an additional half revolution (Figure 6).

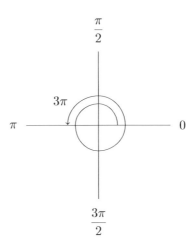

FIGURE 6. The angle of 3π radians.

It is possible for two different angles to terminate at the same location. Consider the scenario given in Figure 7. The two angles $\theta_1 = -\frac{\pi}{2}$ and $\theta_2 = \frac{3\pi}{2}$ are different, yet both end on the vertical axis.

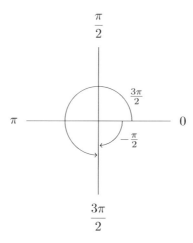

FIGURE 7. Different angles with the same termination location.

Defining the two fundamental trigonometric functions *sine* and *cosine* has two steps. The first step is to define sine and cosine for right triangles. Once we have that definition, we can use it to arrive at a more generalized definition of sine and cosine for all possible angles.

DEFINITION 7.2. Sine and Cosine for Right Triangles
For right triangle $\triangle ABC$

where $0 < A < \frac{\pi}{2}$ we define

$$\text{sine of } \angle A = \frac{\text{length of } \overline{BC}}{\text{length of } \overline{AB}} \qquad \text{cosine of } \angle A = \frac{\text{length of } \overline{AC}}{\text{length of } \overline{AB}}$$

Sine may be abbreviated as "sin" and cosine as "cos".

The right triangle definitions of sine and cosine are applicable to radian values of θ where $0 < \theta < \frac{\pi}{2}$. As our study of trigonometry will involve circles and angles outside of this limited range, this definition is insufficient. We must extend the definitions of sine and cosine to all possible values of θ.

Consider a circle, centered at the origin, of radius 1. Let us label the point $(1, 0)$ as D.

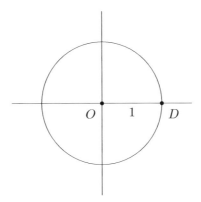

If we draw a line from the origin until it intersects the circle (we will show this intersection happening in the first quadrant only to simplify the following explanation, any quadrant is valid) and then drop a new line straight down to the horizontal axis, we can form a right triangle. The hypotenuse of this right triangle has length 1 (the radius of the circle), and we have labeled angle $\angle AOB$ as θ (Figure 8).

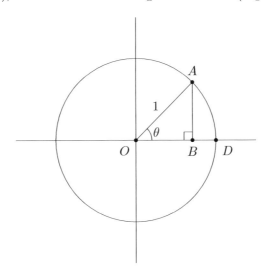

FIGURE 8. Drawing a triangle in quadrant one.

Let us first consider θ. The number of radians in θ is given by

$$\theta = \frac{\text{arc length of } DA}{r} = \frac{\text{arc length of } DA}{1} = \text{arc length of } DA$$

Be sure you understand the significance of this relationship. Because we chose a circle with radius equal to one (also known as a unit circle), the number of radians for any arbitrary angle is equal to the length of the arc resulting from the angle.

Based on the right triangle definitions of sine and cosine we have

$$\sin \theta = \frac{\text{length of } \overline{AB}}{\text{length of } \overline{OA}} = \frac{\text{length of } \overline{AB}}{1} = \text{length of } \overline{AB}$$

$$\cos \theta = \frac{\text{length of } \overline{OB}}{\text{length of } \overline{OA}} = \frac{\text{length of } \overline{OB}}{1} = \text{length of } \overline{OB}$$

Once again, our choice of the unit circle (length of $\overline{OA} = 1$) resulted in some significant simplification.

Finally, consider the two ways in which we can arrive at point A. One way would be to plot point A using the traditional coordinate plotting approach. The coordinates of point A are

$$\left(\text{length of } \overline{OB}, \text{length of } \overline{AB}\right) = (\cos \theta, \sin \theta)$$

Alternatively, we can arrive at point A by starting at the point $(1, 0)$ and moving counterclockwise along the unit circle a distance of θ. The endpoint of this arc with length θ is point A. The two different ways in which we can arrive at point A are the basis for our generalized definition of sine and cosine.

DEFINITION 7.3. Sine and Cosine

For any real value of θ, start at the point $(1, 0)$ on the unit circle. Move around the circle a distance of $|\theta|$. For $\theta > 0$, move in a counter-clockwise direction. For $\theta < 0$, move in a clockwise direction.

$$\sin \theta \text{ is the vertical coordinate of the endpoint}$$

$$\cos \theta \text{ is the horizontal coordinate of the endpoint}$$

Both sine and cosine satisfy the definition of a function. We will explore this fact extensively in upcoming sections. For now, knowing that these trigonometric definitions are also functions will be sufficient.

We will end this section with the first of many relationships between sine and cosine. You may recall that all points on a circle of radius 1, centered at the origin, satisfy the equation

$$x^2 + y^2 = 1$$

As we have just defined $\sin\theta$ and $\cos\theta$ to be the vertical and horizontal coordinates for all points on the unit circle, we can substitute $\cos\theta$ for x and $\sin\theta$ in for y and arrive at the following relationship

$$(\cos\theta)^2 + (\sin\theta)^2 = 1$$

All trigonometric functions have some unique exponential notation that you must become familiar with. The exponents in the previous equation can be written in one additional (and extremely common) manner. Rather than writing the exponent outside a set of parentheses after the trigonometric function, it is possible to write the exponent after the name of the trigonometric function. Our previous equation then becomes

$$\cos^2\theta + \sin^2\theta = 1$$

With either notation, the mathematical operations are the same.

$$\cos^2\theta = (\cos\theta)^2 \quad \text{and} \quad \sin^2\theta = (\sin\theta)^2$$

$$\cos\theta^2 = \cos(\theta^2) \neq (\cos\theta)^2 \quad \text{and} \quad \sin\theta^2 = \sin(\theta^2) \neq (\sin\theta)^2$$

The final version of this relationship between sine and cosine is given by

$$\sin^2\theta + \cos^2\theta = 1$$

Exercises for Section 7.1.

(1) The following values of θ all correspond to arcs on the unit circle. Sketch a picture illustrating the path of the arc and the approximate endpoint. What are the angle values that correspond to these arcs?

(a) $\theta = \dfrac{\pi}{4}$

(b) $\theta = -\dfrac{\pi}{7}$

(c) $\theta = \dfrac{5}{2}$

(d) $\theta = \dfrac{2\pi}{5}$

(e) $\theta = 4\pi$

(f) $\theta = -\dfrac{\pi}{3}$

(g) $\theta = -\dfrac{8\pi}{5}$ (h) $\theta = -\dfrac{2\pi}{11}$ (i) $\theta = \dfrac{7\pi}{4}$

(2) Using the general definition of sine and cosine:

(a) Determine if sine and cosine are positive or negative in each of the four quadrants of the coordinate plane.

(b) Determine the values of sine and cosine for the following values of θ.

(i) $\theta = 0$ (iii) $\theta = \pi$ (v) $\theta = 2\pi$ (vii) $\theta = -\pi$

(ii) $\theta = -\dfrac{\pi}{2}$ (iv) $\theta = \dfrac{3\pi}{2}$ (vi) $\theta = \dfrac{\pi}{2}$ (viii) $\theta = -\dfrac{3\pi}{2}$

(3) Let θ be an arc that terminates in the second quadrant with $\cos\theta = -0.35$. Find the value of $\sin\theta$. Sketch a picture of the arc, and the endpoint.

(4) Let θ be an arc with $\sin\theta = -0.44$. Find the value of $\cos\theta$. Sketch a picture of the arc and the endpoint.

(5) Suppose you are walking on a circular path. You have traveled exactly halfway around the path, a distance of 5.3 km. If you continue walking along the path for an additional 1.2 km, how many radians would correspond to the full arc you traveled?

(6) You own a successful piano delivery company. One of your customers needs you to deliver ten pianos to their business. The loading dock at their business has a height of 101 cm. You must build a ramp from the ground to the top of the loading dock. The boards you have to build the ramp are 3.5 meters long. Assume the ground is level.

(a) What is the distance between the start of the ramp and the base of the loading dock?

(b) What is the sine of the angle between the ground and the ramp?

(c) Suppose the loading dock height and the length of the boards are doubled. Redo the previous parts of this problem with these new measurements and discuss your results.

7.2. The Unit Circle

This section will illustrate some common arcs and their corresponding endpoint coordinates on the unit circle. To begin, recall the special right triangles.

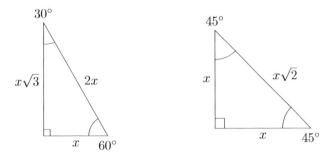

FIGURE 9. The special right triangles

Our goal is to inscribe these triangles into the unit circle. The first step will be to convert all of the angles associated with the special right triangles into radians. Doing so leaves us with

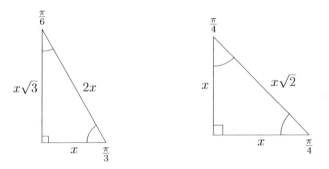

FIGURE 10. The special right triangles (in radians)

As we will be inscribing these triangles into a circle or radius 1, we know that the hypotenuse of each triangle (the longest edge) must have length equal to 1. Setting each hypotenuse length equal to 1 will allow us to solve for x, and then for all other lengths in both triangles. Note that the final step in solving for x in the 45-45-90 ($\frac{\pi}{4}$-$\frac{\pi}{4}$-$\frac{\pi}{2}$ in radians) was to rationalize the denominator.

$$2x = 1 \qquad\qquad x\sqrt{2} = 1$$

$$x = \frac{1}{2} \qquad\qquad x = \frac{1}{\sqrt{2}} = \frac{\sqrt{2}}{2}$$

Plugging these respective values into the corresponding special right triangle leaves us with the following

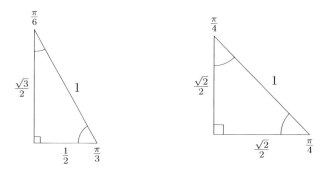

FIGURE 11. The special right triangles, in radians, with hypotenuse equal to 1

Let us begin by inscribing the 45-45-90 ($\frac{\pi}{4}$-$\frac{\pi}{4}$-$\frac{\pi}{2}$) triangle into the first quadrant of the unit circle.

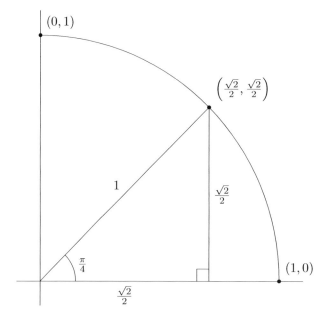

The point of intersection between the triangle and the circle has coordinates $\left(\frac{\sqrt{2}}{2}, \frac{\sqrt{2}}{2}\right)$. Because we are on the unit circle using radians, we know that the arc length from the point $(1,0)$ to the point $\left(\frac{\sqrt{2}}{2}, \frac{\sqrt{2}}{2}\right)$ must be $\frac{\pi}{4}$. And, given that we know this arc length, we can use the definitions of sine and cosine. The definitions tell us that

$$\sin\frac{\pi}{4} = \frac{\sqrt{2}}{2} \quad \text{and} \quad \cos\frac{\pi}{4} = \frac{\sqrt{2}}{2}$$

The process of inscribing the 30-60-90 ($\frac{\pi}{6}$-$\frac{\pi}{3}$-$\frac{\pi}{2}$ in radians) right triangle is similar. However, there are now two possible orientations of this special right triangle (as both of the non-right angles are different). Both orientations are shown below.

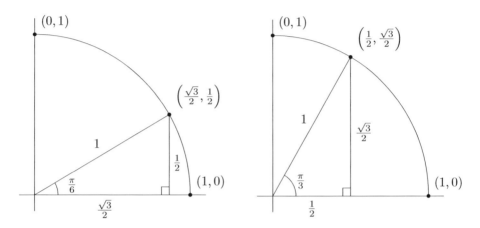

These figures (along with the definitions of sine and cosine) allow us to write out the following relationships

$$\sin\frac{\pi}{6} = \frac{1}{2} \qquad\qquad \cos\frac{\pi}{6} = \frac{\sqrt{3}}{2}$$

$$\sin\frac{\pi}{3} = \frac{\sqrt{3}}{2} \qquad\qquad \cos\frac{\pi}{3} = \frac{1}{2}$$

The definitions of sine and cosine are valid for an arc length of 0, and an arc length of $\frac{\pi}{2}$. These arc lengths terminate at the points $(1,0)$ and $(0,1)$ respectively. Applying the definitions of sine and cosine would give us

$$\sin 0 = 0 \qquad\qquad \cos 0 = 1$$

$$\sin\frac{\pi}{2} = 1 \qquad\qquad \cos\frac{\pi}{2} = 0$$

This completes the first quadrant of the unit circle. A summary of all discussed arc lengths and endpoint coordinates is given in Figure 13.

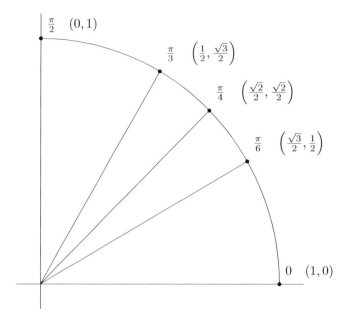

FIGURE 13. First quadrant of the unit circle.

The result of the first quadrant can be extended into the third quadrant. Consider the $\frac{\pi}{6}$ arc length and its endpoint coordinate $\left(\frac{\sqrt{3}}{2}, \frac{1}{2}\right)$. Suppose we extend the line connecting the points $(0,0)$ and $\left(\frac{\sqrt{3}}{2}, \frac{1}{2}\right)$ into the third quadrant until it intersects the unit circle at point A, shown in Figure 14.

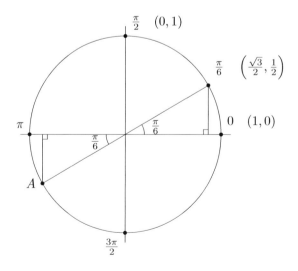

FIGURE 14. Extension into the third quadrant.

The intersection of this line and the horizontal axis creates a pair of vertical angles. If we draw a vertical line from point A to the horizontal axis, we can see the resulting triangle is a 30-60-90 ($\frac{\pi}{6}$-$\frac{\pi}{3}$-$\frac{\pi}{2}$).

We know that the distance from $(0,0)$ to point A is equal to one (the radius of the unit circle). Since both of these triangles share all of the same angles, this means these two triangles are identical (albeit at a different orientation). Therefore the coordinates of point A must be $\left(-\frac{\sqrt{3}}{2}, -\frac{1}{2}\right)$, and the arc length from $(1,0)$ to point A must be $\pi + \frac{\pi}{6} = \frac{7\pi}{6}$. This process can be repeated for both of the $\frac{\pi}{4}$ and $\frac{\pi}{3}$ arc lengths. We leave that as an exercise, and show the results in Figure 15 below.

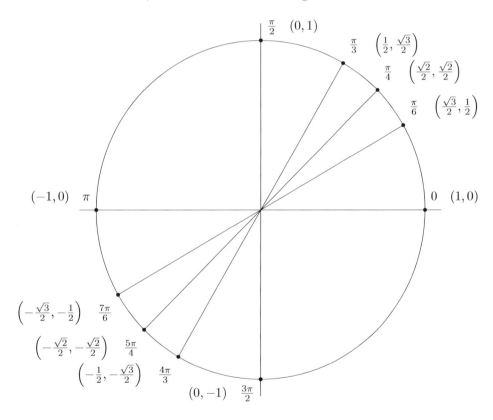

FIGURE 15. First and third quadrants of the unit circlet .

The third quadrant arc lengths and endpoint coordinates could also be the result of the same special right triangles that we used in the first quadrant, but with a different orientation. Let us take this idea and apply it to quadrant two (Figure 16).

Once again, all of the distances for the sides of the triangles will be the same, only the directions will have changed. This fact allows us to calculate the intersection points between the triangles and the unit circle. The arc lengths come from subtracting each associated inner triangle angle from π. Consider the endpoint coordinate of $\left(-\frac{\sqrt{3}}{2}, \frac{1}{2}\right)$. We know that the arc length associated with this endpoint is exactly $\frac{\pi}{6}$ radians short of π radians (as the inner angle on the triangle is $\frac{\pi}{6}$).

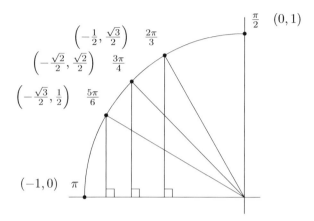

FIGURE 16. Special right triangles in the second quadrant of the unit circle.

The same process of reorienting the special right triangles into the fourth quadrant will allow the calculation of the corresponding intersection points. Since all lengths are the same, only the directions will change. The associated arc lengths are the result of subtracting the inner triangle angles from 2π (Figure 17).

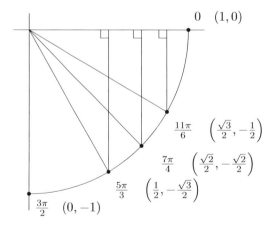

FIGURE 17. Special right triangles in the fourth quadrant of the unit circle.

The full unit circle is given in Figure 18. These particular arc length and endpoint combinations are extremely common throughout trigonometry and beyond. Knowledge of these arc lengths and endpoints (be it through construction with the right triangles or memorization) will become critical in the very near future. Equally as important will be the use of sine and cosine on the unit circle. The sine of any arc on the unit circle is the vertical coordinate of the endpoint. Likewise, the cosine of any arc on the unit circle is the horizontal coordinate of the endpoint.

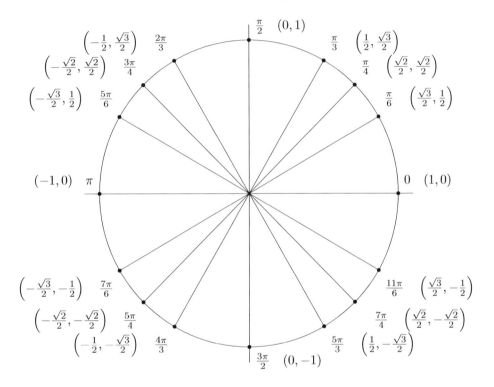

FIGURE 18. The unit circle.

Exercises for Section 7.2.

(1) Sketch the full unit circle, with all labeled endpoints. Do not consult any reference material.

(2) Evaluate each of the following using exact values. You should evaluate these mentally, not by copying down answers from the previous sketch.

(a) $\sin\left(\dfrac{3\pi}{4}\right)$

(b) $\cos\left(\dfrac{4\pi}{3}\right)$

(c) $\sin(7\pi)$

(d) $\sin(\pi)$

(e) $\cos\left(\dfrac{\pi}{3}\right)$

(f) $\sin\left(-\dfrac{5\pi}{6}\right)$

(g) $\cos\left(-\dfrac{11\pi}{6}\right)$

(h) $\cos(-3\pi)$

(i) $\cos\left(\dfrac{3\pi}{2}\right)$

(j) $\cos(92\pi)$

(k) $\sin\left(-\dfrac{5\pi}{4}\right)$

(l) $\sin\left(\dfrac{5\pi}{3}\right)$

(3) Write a ten question quiz that you feel would be a good evaluation for general knowledge of the unit circle. You may not use any of the values in the previous question. Be sure to include an answer key.

7.3. Combinations of sine and cosine

This section will explore relationships of sine and cosine. We begin by considering an arbitrary arc on the unit circle, call it θ. The endpoints of θ and $-\theta$ are shown below. Our work in the previous sections tells us that the endpoint coordinates of θ are

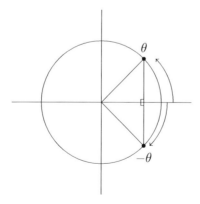

$(\cos\theta, \sin\theta)$, while $-\theta$ has endpoint coordinates $(\cos(-\theta), \sin(-\theta))$. The horizontal coordinates are identical, while the vertical coordinates are opposites. Put another way, we have

$$\cos\theta = \cos(-\theta) \quad \text{and} \quad \sin(-\theta) = -\sin\theta$$

Our earlier work in function theory tells us that this relationship qualifies cosine as an even function, while sine is an odd function. Take a moment to verify that cosine is odd, and sine is even, regardless of the quadrant in which θ happens to terminate.

Algebraic combinations of sine and cosine give rise to four other trigonometric functions. Tangent, cotangent, secant and cosecant are defined as follows

$$\text{Tangent: } \tan\theta = \frac{\sin\theta}{\cos\theta} \qquad \text{Cotangent: } \cot\theta = \frac{\cos\theta}{\sin\theta}$$

$$\text{Secant: } \sec\theta = \frac{1}{\cos\theta} \qquad \text{Cosecant: } \csc\theta = \frac{1}{\sin\theta}$$

As these additional trigonometric functions are all defined in terms of sine and cosine, we can use the unit circle to evaluate them. Suppose we were interested in the exact value of $\tan\left(\frac{\pi}{3}\right)$.

$$\tan\left(\frac{\pi}{3}\right) = \frac{\sin\left(\frac{\pi}{3}\right)}{\cos\left(\frac{\pi}{3}\right)} = \frac{\frac{\sqrt{3}}{2}}{\frac{1}{2}} = \frac{\sqrt{3}}{2} \cdot \frac{2}{1} = \frac{\sqrt{3}}{1} = \sqrt{3}$$

Similarly, something like $\csc\left(\frac{5\pi}{4}\right)$ would be

$$\csc\left(\frac{5\pi}{4}\right) = \frac{1}{\sin\left(\frac{5\pi}{4}\right)} = \frac{1}{-\frac{\sqrt{2}}{2}} = -\frac{2}{\sqrt{2}} = -\sqrt{2}$$

Given that we know the symmetry of sine and cosine (odd and even, respectively), we can consider the symmetry of the additional trigonometric functions as well. The symmetry of tangent is given by

$$\tan(-\theta) = \frac{\sin(-\theta)}{\cos(-\theta)} = \frac{-\sin\theta}{\cos\theta} = -\frac{\sin\theta}{\cos\theta} = -\tan\theta$$

which tells us that tangent is an odd function.

Now, let us put all of our knowledge of the unit circle to use with some examples. Suppose an arc of length θ terminates in the second quadrant, and that $\sin\theta = 0.53$. Find $\sin(\pi - \theta)$, $\cos\theta$, and $\tan(-\theta)$.

To identify a rough placement of the endpoint for arc length θ, start at the point $(1, 0)$ and move counter-clockwise around the unit circle a distance of θ. We know that θ terminates in the second quadrant, and as $\sin\theta = 0.53$, we know that the vertical coordinate of the endpoint is 0.53.

An upcoming discussion will cover how to properly evaluate trigonometric addition of the form $\sin(\alpha \pm \beta)$, but an intuitive approach will serve for this specific example.

To identify the endpoint of $\pi - \theta$, start at the endpoint for an arc of length π, the point $(-1, 0)$. Then, move in the clockwise (negative) direction a distance of θ. Note that the arc length between the endpoint of arc θ and the point $(-1, 0)$ will be the same as the arc length between $(1, 0)$ to the endpoint of arc $\pi - \theta$. As $\sin \theta = 0.53$, we know that the vertical coordinate for the endpoint of $\pi - \theta$ must also be 0.53.

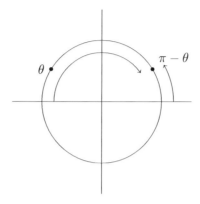

Evaluation of $\cos \theta$ can be achieved through the identity $\sin^2 \theta + \cos^2 \theta = 1$.

$$(0.53)^2 + \cos^2 \theta = 1$$

$$\cos^2 \theta = 1 - (0.53)^2$$

$$\cos \theta = \pm\sqrt{1 - (0.53)^2}$$

As θ terminated in quadrant two, we know that $\cos \theta$ (the horizontal coordinate of the endpoint) must be negative. Thus, we have

$$\cos \theta = -\sqrt{1 - (0.53)^2} \approx -0.848$$

Evaluation of $\tan(-\theta)$ is given by

$$\tan(-\theta) = \frac{\sin(-\theta)}{\cos(-\theta)} = \frac{-\sin \theta}{\cos \theta} = \frac{-0.53}{-\sqrt{1 - (0.53)^2}} \approx 0.625$$

For another example, suppose $\sin \theta = \frac{5}{13}$, and we wish to find $\cos \theta$. This problem is less specific than the previous, as we do not know in which quadrant θ terminates. We do know the vertical coordinate of the endpoint ($\sin \theta$) is $\frac{5}{13}$, so θ must terminate in either quadrant one or quadrant two. Even if we do not know the exact location of

7.3. COMBINATIONS OF SINE AND COSINE

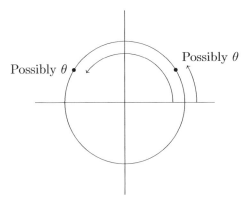

the endpoint, the relationship of $\sin^2 \theta + \cos^2 \theta = 1$ still holds.

$$\left(\frac{5}{13}\right)^2 + \cos^2 \theta = 1$$

$$\cos^2 \theta = 1 - \left(\frac{5}{13}\right)^2$$

$$\cos \theta = \pm\sqrt{1 - \left(\frac{5}{13}\right)^2}$$

As θ can potentially terminate in either quadrant one or quadrant two, we require both the plus and the minus answer for cosine. Termination in quadrant one would result in the positive answer, while termination in quadrant two would result in the negative answer.

$$\cos \theta = \pm\sqrt{1 - \left(\frac{5}{13}\right)^2} \approx \pm 0.923$$

Exercises for Section 7.3.

(1) Evaluate the tangent, secant, and cosecant of the following values. Your answers should be exact (non-decimal).

(a) $\theta = \pi$

(b) $\theta = \dfrac{2\pi}{3}$

(c) $\theta = \dfrac{\pi}{2}$

(d) $\theta = 0$

(e) $\theta = \dfrac{5\pi}{4}$

(f) $\theta = \dfrac{4\pi}{3}$

(g) $\theta = \dfrac{11\pi}{2}$

(h) $\theta = \dfrac{15\pi}{4}$

(2) Sketch three different scenarios on the unit circle. The first sketch should have an arc, θ, that terminates in quadrant two. The second should have θ terminate in quadrant three, and the third in quadrant four. For each of the three sketches, illustrate that $\cos(\theta) = \cos(-\theta)$ and $\sin(-\theta) = -\sin(\theta)$.

(3) Use algebra to determine if the functions of secant, cosecant, and cotangent are even, odd, or neither.

(4) Suppose you are working through a trigonometry exercise and arrive at the result

$$\sin \theta = 1.2$$

Explain why this is impossible.

(5) Let θ terminate in quadrant two and $\cos \theta = -0.62$. Evaluate each of the following.

(a) $\sin \theta$

(b) $\tan \theta$

(c) $\sec \theta$

(d) $\csc \theta$

(e) $\cos(\pi - \theta)$

(f) $\cos(\theta + \pi)$

(g) $\sin(\pi - \theta)$

(h) $\sin(\theta + \pi)$

7.4. Graphing Sine and Cosine

Thus far, our work in trigonometry has largely focused on definitions. The time has come to begin exploring graphs of trigonometric functions. Further still, we can finally begin using function notation with our trigonometric definitions. Consider a function f where

$$f(x) = \sin x$$

and x is in radians. Our work with the unit circle has already provided us with sixteen points on the graph of this function. A traditional table view is below.

x	$\sin x$	x	$\sin x$	x	$\sin x$	x	$\sin x$
0	0	$\frac{\pi}{2}$	1	π	0	$\frac{3\pi}{2}$	-1
$\frac{\pi}{6}$	$\frac{1}{2}$	$\frac{2\pi}{3}$	$\frac{\sqrt{3}}{2}$	$\frac{7\pi}{6}$	$-\frac{1}{2}$	$\frac{5\pi}{3}$	$-\frac{\sqrt{3}}{2}$
$\frac{\pi}{4}$	$\frac{\sqrt{2}}{2}$	$\frac{3\pi}{4}$	$\frac{\sqrt{2}}{2}$	$\frac{5\pi}{4}$	$-\frac{\sqrt{2}}{2}$	$\frac{7\pi}{4}$	$-\frac{\sqrt{2}}{2}$
$\frac{\pi}{3}$	$\frac{\sqrt{3}}{2}$	$\frac{5\pi}{6}$	$\frac{1}{2}$	$\frac{4\pi}{3}$	$-\frac{\sqrt{3}}{2}$	$\frac{11\pi}{6}$	$-\frac{1}{2}$

Plotting these points would produce the following graph.

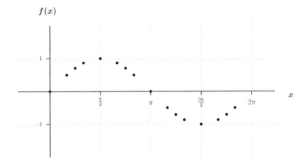

FIGURE 19. The values of $f(x) = \sin x$ from the unit circle

Our definitions of the functions sine and cosine will allow any real radian value as input. In other words, the full graph of $f(x) = \sin x$ is not limited to the 16 points shown above, but is instead defined for all real values of x. Figure 20 illustrates the full graph of the sine function.

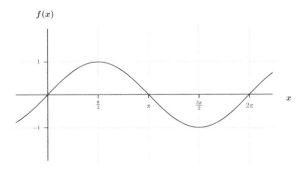

FIGURE 20. The function $f(x) = \sin x$

The sine function continues indefinitely for all positive and negative values of x, and is *periodic* (repeats itself). The *period* of the sine function (how long it takes for the function to repeat itself) is 2π radians.

The *amplitude* of the sine function can be described in two equivalent ways. One way would be to consider the amplitude as the maximum vertical distance away from *equilibrium* (Figure 20 has an equilibrium value of 0). The second way to consider the amplitude would be

$$\text{Amplitude} = \frac{\text{Maximum vertical value} - \text{Minumum vertical value}}{2}$$

Either approach will yield the same answer, in this case a value of 1.

Turning points are points in which the function changes from increasing to decreasing, or vice versa. The sine function has turning points when

$$x = \frac{\pi}{2} + k\pi$$

where $k \in \mathbb{Z}$. Note that the notation of $k\pi$ is simply a compact way of adding or subtracting any multiple of π.

Inflection points are points where the concavity of the function changes from concave up to concave down, or vice versa. The sine function has inflection points when $x = k\pi$ where $k \in \mathbb{Z}$.

The graph of

$$f(x) = \cos x$$

is given in Figure 21. Similar to the sine function, cosine also has a period of 2π, and an amplitude of 1. Unlike the sine function, cosine has turning points when $x = k\pi$ where $k \in \mathbb{Z}$, and inflection points when $x = \frac{\pi}{2} + k\pi$ where $k \in \mathbb{Z}$.

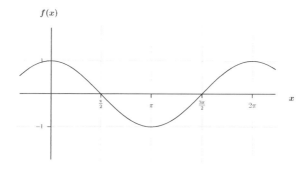

FIGURE 21. The function $f(x) = \cos x$

Note the similarities between the graphs of sine and cosine. A future discussion will deal with trigonometric transformations in detail, but for the moment consider what might happen if we shift all points on the sine graph to the left a distance of $\frac{\pi}{2}$. Visually, it seems this would result in the cosine function. This is correct. A horizontal transformation on the sine function of $\frac{\pi}{2}$ to the left would result in the cosine function. Written out algebraically, we would have

$$\sin\left(x + \frac{\pi}{2}\right) = \cos x$$

Similarly, consider what might be the result of something like

$$\cos\left(\frac{\pi}{2} - x\right)$$

As written, these two transformations on the cosine function would be

1. Shift left by $\frac{\pi}{2}$

2. Reflect across the vertical axis

Spend a moment considering the result of these two transformations. It may help to draw several periods of the cosine function on both sides of $x = 0$. Hopefully, at the end of your consideration, we can agree that

$$\cos\left(\frac{\pi}{2} - x\right) = \sin x$$

A similar series of arguments gives the relationship

$$\sin\left(\frac{\pi}{2} - x\right) = \cos x$$

While we could continue this exercise of exploration indefinitely, a considerably better use of our time would be to determine an algebraic formula for

$$\sin(\alpha \pm \beta) \quad \text{and} \quad \cos(\alpha \pm \beta)$$

where α and β represent arbitrary arcs on the unit circle. As is often the case in trigonometry, we approach this task through use of the unit circle.

Suppose we have two different arcs on the unit circle. Call them α and β. The coordinates at the endpoint of α are $(\cos \alpha, \sin \alpha)$, while the coordinates at the endpoint of β are $(\cos \beta, \sin \beta)$. Let d represent the linear distance between these endpoints.

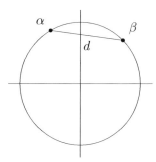

FIGURE 22. Arcs α and β, and d, the linear distance between them.

If β is subtracted from both of these arc lengths, the arc $\alpha - \beta$ would have endpoint coordinates $(\cos(\alpha - \beta), \sin(\alpha - \beta))$, while $\beta - \beta$ would move back to the starting location of $(1, 0)$. The length of d remains the same.

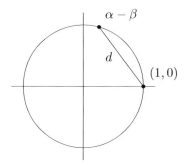

FIGURE 23. Arc $\alpha - \beta$, and d, the linear distance between the arc endpoint and $(1, 0)$.

Recall that the distance between any two points in a coordinate plane is given by

$$\text{distance} = \sqrt{(x_2 - x_1)^2 + (y_2 - y_1)^2}$$

Each of the previous figures provided a different way to represent the same distance, d. From the first we have

$$d = \sqrt{(\cos \alpha - \cos \beta)^2 + (\sin \alpha - \sin \beta)^2}$$

while the second image gives us

$$d = \sqrt{(\cos(\alpha - \beta) - 1)^2 + (\sin(\alpha - \beta) - 0)^2}$$

As d is the same in both cases, it must be true that

$$\sqrt{(\cos\alpha - \cos\beta)^2 + (\sin\alpha - \sin\beta)^2} = \sqrt{(\cos(\alpha - \beta) - 1)^2 + (\sin(\alpha - \beta) - 0)^2}$$

Squaring both sides would gives us an equality that we will refer to as equation 7.1.

(7.1) $\quad (\cos\alpha - \cos\beta)^2 + (\sin\alpha - \sin\beta)^2 = (\cos(\alpha - \beta) - 1)^2 + (\sin(\alpha - \beta) - 0)^2$

We can simplify this equation considerably. To avoid confusion, we begin by working only with the left side of equation 7.1. Note that

$$(\cos\alpha - \cos\beta)^2 = \cos^2\alpha - 2\cos\alpha\cos\beta + \cos^2\beta$$

$$(\sin\alpha - \sin\beta)^2 = \sin^2\alpha - 2\sin\alpha\sin\beta + \sin^2\beta$$

Making these substitutions into the left side of the equation 7.1 gives us

$$\cos^2\alpha - 2\cos\alpha\cos\beta + \cos^2\beta + \sin^2\alpha - 2\sin\alpha\sin\beta + \sin^2\beta$$

which can be rewritten as

$$\left(\cos^2\alpha + \sin^2\alpha\right) + \left(\cos^2\beta + \sin^2\beta\right) - 2\cos\alpha\cos\beta - 2\sin\alpha\sin\beta$$

and then simplified to

$$1 + 1 - 2\cos\alpha\cos\beta - 2\sin\alpha\sin\beta = 2 - 2\left(\cos\alpha\cos\beta + \sin\alpha\sin\beta\right)$$

Pausing work on the left side for a moment, let us consider the right side of equation 7.1.

$$(\cos(\alpha - \beta) - 1)^2 + (\sin(\alpha - \beta) - 0)^2$$

Expanding this out would give

$$\cos^2(\alpha - \beta) - 2\cos(\alpha - \beta) + 1 + \sin^2(\alpha - \beta)$$

regrouping terms leaves us with

$$\left(\cos^2(\alpha - \beta) + \sin^2(\alpha - \beta)\right) - 2\cos(\alpha - \beta) + 1$$

which can be simplified to

$$1 - 2\cos(\alpha - \beta) + 1 = 2 - 2\cos(\alpha - \beta)$$

With both sides of equation 7.1 simplified, we can once again set them equal to one another

$$2 - 2(\cos\alpha\cos\beta + \sin\alpha\sin\beta) = 2 - 2\cos(\alpha - \beta)$$

Some simple cancellation reveals our first trigonometric addition formula

(7.2) $$\cos\alpha\cos\beta + \sin\alpha\sin\beta = \cos(\alpha - \beta)$$

As α and β are arbitrary arcs of our choosing, we can replace β in equation 7.2 with $-\beta$.

$$\cos\alpha\cos(-\beta) + \sin\alpha\sin(-\beta) = \cos(\alpha - (-\beta))$$

Simplification brings us to our second trigonometric addition formula

(7.3) $$\cos\alpha\cos\beta - \sin\alpha\sin\beta = \cos(\alpha + \beta)$$

Once again, note that α and β are arbitrary arcs on the unit circle. Suppose we replace α in equation 7.3 with $\frac{\pi}{2} - \alpha$.

$$\cos\left(\frac{\pi}{2} - \alpha\right)\cos\beta - \sin\left(\frac{\pi}{2} - \alpha\right)\sin\beta = \cos\left(\frac{\pi}{2} - \alpha + \beta\right)$$

Substitution of $\cos\left(\frac{\pi}{2} - \alpha\right) = \sin\alpha$ and $\sin\left(\frac{\pi}{2} - \alpha\right) = \cos\alpha$ on the left hand side gives us

$$\sin\alpha\cos\beta - \cos\alpha\sin\beta = \cos\left(\frac{\pi}{2} - (\alpha - \beta)\right)$$

and a final substitution of $\cos\left(\frac{\pi}{2} - (\alpha - \beta)\right) = \sin(\alpha - \beta)$ on the right hand side bring us to our third trigonometric addition formula

(7.4) $$\sin\alpha\cos\beta - \cos\alpha\sin\beta = \sin(\alpha - \beta)$$

Lastly, if we replace β in equation 7.4 with $-\beta$ we have

$$\sin\alpha\cos(-\beta) - \cos\alpha\sin(-\beta) = \sin(\alpha - (-\beta))$$

which simplifies to our final addition formula

(7.5) $$\sin\alpha\cos\beta + \cos\alpha\sin\beta = \sin(\alpha + \beta)$$

The trigonometric addition formulas are powerful tools. While it is not necessary for you to be able to derive them, you must know them and be comfortable using them. All of the addition formula are presented below.

$$\sin(\alpha + \beta) = \sin\alpha\cos\beta + \cos\alpha\sin\beta$$
$$\sin(\alpha - \beta) = \sin\alpha\cos\beta - \cos\alpha\sin\beta$$
$$\cos(\alpha + \beta) = \cos\alpha\cos\beta - \sin\alpha\sin\beta$$
$$\cos(\alpha - \beta) = \cos\alpha\cos\beta + \sin\alpha\sin\beta$$

Exercises for Section 7.4.

(1) Discuss why the the definitions of sine and cosine satisfy the definition of a function. Part of your discussion should consider what it would mean if sine and cosine did not satisfy the definition of a function.

(2) Sketch graphs of

$$f(x) = \sin x \quad \text{and} \quad g(x) = \cos x$$

on the interval $[-\pi, \pi]$. Label all points that correspond to the sixteen common points on the unit circle. Also label zeros, turning points, and inflection points.

(3) Evaluate each of the following. Your answers should be exact (non-decimal).

(a) $\sin\left(\dfrac{4\pi}{3} + \dfrac{\pi}{4}\right)$ (b) $\cos\left(\dfrac{\pi}{6} - \dfrac{\pi}{3}\right)$ (c) $\tan\left(\dfrac{7\pi}{4} - \dfrac{2\pi}{3}\right)$

(d) $\sec\left(\dfrac{\pi}{3} + \dfrac{5\pi}{6}\right)$ (e) $\csc\left(\dfrac{3\pi}{4} + \dfrac{11\pi}{6}\right)$ (f) $\cot\left(\dfrac{7\pi}{12}\right)$

(4) Work through the derivation of the trigonometric addition formulas. Focus on any of the steps you do not understand. Discuss anything you find challenging or unclear.

7.5. Graphing Tangent and the Rest

Rather than presenting the graph of tangent, let us first discuss features that we expect the graph to have. Recall that the definition of tangent is

$$f(x) = \tan x = \frac{\sin x}{\cos x}$$

Thinking back to our earlier work with functions, we know interesting behavior will occur when either the numerator or the denominator of this fraction are equal to zero. Consider what will happen when $\cos x = 0$. This would result in division by zero, which immediately tells us that the tangent is not defined at all values of x such that $\cos x = 0$. The values of x that make the cosine function equal to zero are given by

$$\{x \mid x = \frac{\pi}{2} + k\pi \text{ where } k \in \mathbb{Z}\}$$

Furthermore, we expect vertical asymptotes at all of these undefined points.

If $\sin x = 0$, then $\frac{\sin x}{\cos x} = 0$. This tells us that the tangent function has the same zeros as the sine function, namely

$$\{x \mid x = k\pi \text{ where } k \in \mathbb{Z}\}$$

Sine, cosine, tangent's asymptotes and zeros are plotted together in Figure 24.

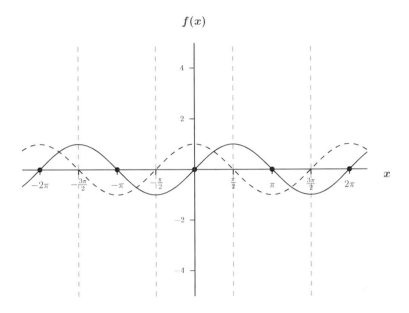

FIGURE 24. Sine (solid) and cosine (dotted) together with the zeros and asymptotes of $f(x) = \tan x$

With the zeros and asymptotes taken care of, we can turn our attention to what might happen between these asymptotes. Rather than try to visualize the entire graph at once, let us focus on domain values between $-\frac{\pi}{2}$ and , $\frac{\pi}{2}$. As sine and cosine approach $\frac{\pi}{2}$ from the left, cosine is positive and decreasing to zero, while sine is positive and increasing to one. Tangent is undefined at $\frac{\pi}{2}$, so sine will never reach a value of one and cosine will never reach a value of zero, but they can each get as close as we want them to. This intuitively makes the value of tangent

$$\frac{\text{a positive number, less than 1, always getting closer to 1}}{\text{a positive number, less than 1, always getting closer to 0}}$$

Experiment with this idea on a calculator. You should notice that as the denominator decreases towards zero, while the numerator increases towards one, the fraction overall is increasing drastically. Effectively, the fraction can increase without limit. Whenever we might need to make it larger all we need do is make the denominator smaller, and the numerator closer to one. Thus, as x approaches $\frac{\pi}{2}$ from the left, $\tan x$ increases without bound.

The exact same argument applies as x approaches $-\frac{\pi}{2}$ from the right, the only difference is sine is negative and decreasing towards negative one. Written in the intuitive manner we used earlier, this produces

$$\frac{\text{a negative number, greater than } -1, \text{ always getting closer to } -1}{\text{a positive number, less than 1, always getting closer to 0}}$$

Overall, the fraction that is tangent becomes negative, so tangent decreases without bound. This same approach can be used to reason your way through all of tangent's behavior as it approaches any and all of its asymptotes (and zeros). The full graph of the tangent function is given in Figure 25.

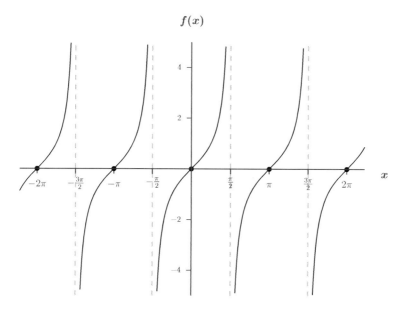

FIGURE 25. $f(x) = \tan x$

As it has no maximum or minimum value, amplitude is not defined for the tangent function. It is visually clear that tangent has a period of π, and this fact can be verified algebraically as well.

$$\tan(x + \pi) = \frac{\sin(x + \pi)}{\cos(x + \pi)} = \frac{\sin x \cos \pi + \sin \pi \cos x}{\cos x \cos \pi - \sin x \sin \pi} = \frac{(\sin x)(-1)}{(\cos x)(-1)} = \frac{\sin x}{\cos x} = \tan x$$

Unlike sine and cosine, tangent is not defined on all of $\mathbb{R}$. Instead, the domain of the tangent function is

$$\{x \mid x \in \mathbb{R} \text{ and } x \neq \frac{\pi}{2} + k\pi \text{ where } k \in \mathbb{Z}\}$$

The process that was used in graphing tangent will work for cotangent, secant, and cosecant. We leave these graphs as exercises.

Lastly, consider once again that the definition of the tangent function is given by

$$f(x) = \tan x = \frac{\sin x}{\cos x}$$

and recall that the output of sine and cosine are (respectively) the vertical and horizontal coordinates of the endpoint for an arc of length x on the unit circle. This means that an additional way to interpret the output of the tangent function is as the slope of the line that connects the origin, and the endpoint of arc length x.

Exercises for Section 7.5.

(1) Determine the behavior of

$$f(x) = \sec x \qquad g(x) = \csc x \qquad h(x) = \cot x$$

on the interval of $[-2\pi, 2\pi]$. Follow the ideas that were presented for the development of tangent's graph. Include the location of zeros and asymptotes. Sketch graphs of f, g, and h. Your work for each of these three functions should be as detailed as was the work for the development of tangent's graph.

(2) Given the four distinct arcs

$$\theta_1 = \frac{\pi}{3} \qquad \theta_2 = \frac{3\pi}{4} \qquad \theta_3 = \frac{3\pi}{2} \qquad \theta_4 = \frac{11\pi}{6}$$

sketch four different graphs, each of which depicts the arc on the unit circle. Illustrate how the tangent of the arc is the slope of the line connecting the origin and the arc endpoint.

(3) Let the functions f and g be defined by

$$f(x) = \tan x \qquad \text{and} \qquad g(x) = \frac{1}{\frac{\cos x}{\sin x}}$$

Does $f = g$? Provide reasoning for your conclusion.

7.6. Vertical Transformations of Trigonometric Functions

Transformations of trigonometric functions operate in the same fashion as was described in Chapter 3. Rather than repeat the material presented there, we begin with some examples.

Suppose
$$g(x) = 3\cos x$$

This transformation would be a vertical stretch by a factor of three on the fundamental function $f(x) = \cos x$. Put another way, we have that $g(x) = 3f(x)$. As the vertical values on function g are three times the vertical values on function f, the amplitude of g must be three times the amplitude of f. The functions f and g are plotted together in Figure 26. Note that the equilibrium position of both functions is still zero.

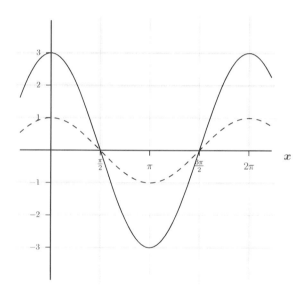

FIGURE 26. $f(x) = \cos x$ (dotted), and $g(x) = 3\cos x$ (solid)

Let
$$g(x) = \cos x - 2$$

This transformation should take the fundamental function $f(x) = \cos x$ and shift down by 2 units. This transformation is illustrated in Figure 27. The amplitudes of f and g are the same, while the maximum, minimum, and equilibrium values have all changed.

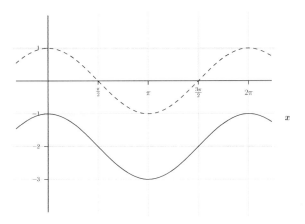

FIGURE 27. $f(x) = \cos x$ (dotted), and $g(x) = \cos x - 2$ (solid)

We have previously seen a graph of the function $f(x) = \tan x$, though the only vertical values we considered were the zeros. Vertical compression (or vertical stretch) of the tangent function is illustrated more effectively if we identify additional vertical values on the graph of tangent. As the definition of tangent is given by

$$\tan x = \frac{\sin x}{\cos x}$$

the tangent will be equal to 1 when $\sin x = \cos x$. The unit circle can provide us with this exact information. Starting at $(1,0)$ and moving clockwise, the first arc such that $\tan x = 1$ is $x = \frac{\pi}{4}$.

$$\tan\left(\frac{\pi}{4}\right) = \frac{\sin\left(\frac{\pi}{4}\right)}{\cos\left(\frac{\pi}{4}\right)} = \frac{\frac{\sqrt{2}}{2}}{\frac{\sqrt{2}}{2}} = 1$$

As tangent has a period of π, we have that $\tan x = 1$ when $x = \frac{\pi}{4} + k\pi$ where $k \in \mathbb{Z}$. Similarly, $\tan x = -1$ when $\sin x = -\cos x$. Examination of the unit circle tells us that this happens when $x = \frac{3\pi}{4} + k\pi$ where $k \in \mathbb{Z}$ (Figure 28).

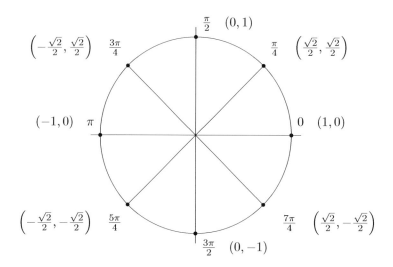

FIGURE 28. Identifying $\tan x = \pm 1$.

We will use these newly identified vertical values to draw a single period of tangent between $\frac{\pi}{2}$ and $\frac{3\pi}{2}$, though any period of tangent could be considered. This figure will be used as the starting graph for transformations of the tangent function.

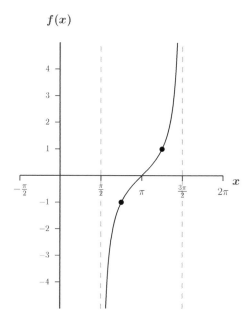

FIGURE 29. A single period of tangent with vertical values of ± 1 labeled

Finally, suppose we would like to graph

$$g(x) = \frac{1}{3}\tan x + 1$$

The function g consists of two transformations on the fundamental function $f(x) = \tan x$. Specifically (and in order) those transformations are a vertical compression by a factor of $\frac{1}{3}$ and a vertical shift up by 1 unit.

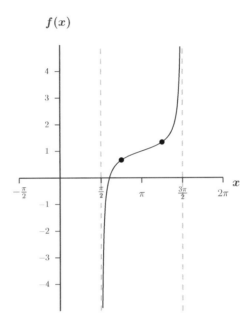

FIGURE 30. $g(x) = \frac{1}{3}\tan x + 1$ The points corresponding to $\left(\frac{3\pi}{4}, -1\right)$ and $\left(\frac{5\pi}{4}, 1\right)$ are shown.

Exercises for Section 7.6.

(1) Each of the following functions are some number of vertical transformations on a fundamental trigonometric function.

(a) $f(x) = 2\sin x - 1$

(b) $g(x) = \frac{1}{3}\cos x + 3$

(c) $h(x) = -\tan x - 2$

(d) $j(x) = 2\csc x - 2$

(e) $k(x) = -\dfrac{1}{2}\sec x$

(g) $m(x) = \dfrac{\pi}{2}\sin x$

(f) $l(x) = \pi \cot x + 1$

(h) $n(x) = \dfrac{2}{\pi}\cos x$

  (i) State the the transformations on a fundamental trigonometric function. Be sure to list them in the correct order.

  (ii) Select three points on the fundamental function and apply the transformations.

  (iii) Use the points to sketch a graph of each transformed function. Your graph should include at least one full period.

  (iv) State the domain and range of the transformed function.

(2) Write the equation of a sine function that oscillates around an equilibrium of e, with amplitude of 3π.

7.7. Horizontal Transformations of Trigonometric Functions

The fundamental movements associated with horizontal transformations of trigonometric functions were presented in Chapter 3. We will begin this section with some examples.

Let the function g be given by

$$g(x) = \sin\left(x - \dfrac{\pi}{2}\right)$$

As we expect, g is a horizontal shift of $\dfrac{\pi}{2}$ to the right of the fundamental function $f(x) = \sin x$. Horizontal shifts of trigonometric functions are referred to as *phase shifts*. In this example, we would say that the function g is the function f with a phase shift of $\dfrac{\pi}{2}$. Both functions are shown in Figure 31.

Now, suppose g makes its pairings according to

$$g(x) = \sin\left(\dfrac{1}{3}x\right)$$

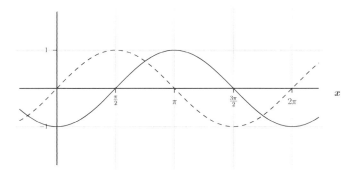

FIGURE 31. $f(x) = \sin x$ (dotted), and $g(x) = \sin\left(x - \frac{\pi}{2}\right)$ (solid)

In this case, the function g is a horizontal stretch by a factor of 3 on the function $f(x) = \sin x$. This stretch will modify the period of g. Specifically, this stretch will mean g has a period three times that of f. As we know the period of the sine function is 2π, the period of function g must be $(2\pi)(3) = 6\pi$.

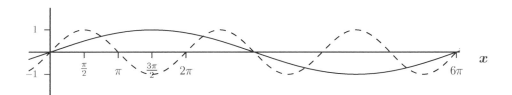

FIGURE 32. $f(x) = \sin x$ (dotted), and $g(x) = \sin\left(\frac{1}{3}x\right)$ (solid)

Multiple horizontal transformations will again result in two equivalent ways to interpret them. Given the function

$$g(x) = \cos\left(2x - \frac{\pi}{6}\right)$$

the two horizontal transformations on $f(x) = \cos x$ are given by either of the following

$$g(x) = \cos\left(2x - \frac{\pi}{6}\right) \qquad\qquad g(x) = \cos\left(2\left(x - \frac{\pi}{12}\right)\right)$$

1. Shift right $\dfrac{\pi}{6}$ 1. Horizontal compression by $\dfrac{1}{2}$

2. Horizontal compression by $\dfrac{1}{2}$ 2. Shift right $\dfrac{\pi}{12}$

Regardless of the choice, the graph of g is given below.

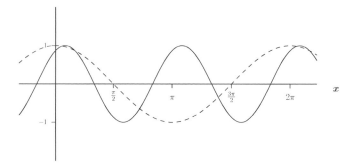

FIGURE 33. $f(x) = \cos x$ (dotted), and $g(x) = \cos\left(2x - \frac{\pi}{6}\right)$ (solid)

Our next exercise will focus on writing trigonometric equations. Specifically, we would like to write equations where we have chosen values for amplitude, phase shift, period, and distance from equilibrium. To illustrate this process, let us write an equation using the sine function such that the graph of the function will have the following visual characteristics; a period of $\frac{\pi}{6}$, amplitude of 7, phase shift of $\frac{\pi}{3}$ and vertical shift from equilibrium of -2.

The vertical transformations on amplitude and equilibrium position can be directly applied to the sine function as

$$f(x) = 7 \sin x - 2$$

For the horizontal, let us first consider the period. Sine has a period of 2π. We can adjust the period to $\frac{\pi}{6}$ through multiplication of a constant value. The algebra below illustrates the process of solving for the unknown constant

$$(2\pi)c = \frac{\pi}{6}$$

$$c = \frac{\pi}{6} \cdot \frac{1}{2\pi} = \frac{1}{12}$$

Put another way, compressing a period of 2π by a factor of $\frac{1}{12}$ will result in a period of $\frac{\pi}{6}$. Keep in mind that the mathematical representation of a horizontal compression by $\frac{1}{12}$ will be written as

$$f(x) = 7 \sin(12x) - 2$$

7.7. HORIZONTAL TRANSFORMATIONS OF TRIGONOMETRIC FUNCTIONS

All that remains is the phase shift of $\frac{\pi}{3}$. It is tempting to include this phase shift as

$$f(x) = 7\sin\left(12x - \frac{\pi}{3}\right) - 2$$

However, that is incorrect. Consider the order of these horizontal transformations as they are written. First, we would shift right $\frac{\pi}{3}$, then we would compress horizontally by a factor of $\frac{1}{12}$. Performing the compression after the shift reduces the shift to $\frac{1}{12} \cdot \frac{\pi}{3} = \frac{\pi}{36}$. To fix this, we must perform these transformations in the other order, or phase shift 12 times further.

$$f(x) = 7\sin\left(12\left(x - \frac{\pi}{3}\right)\right) - 2 = 7\sin\left(12x - 4\pi\right) - 2$$

Either version of f as written above will result in an equation of the sine function with the desired characteristics.

Exercises for Section 7.7.

(1) Each of the following functions are some combination of vertical and horizontal transformations on a fundamental trigonometric function.

(a) $f(x) = \frac{2}{7}\sin\left(x - \frac{\pi}{3}\right)$

(b) $g(x) = \pi\cos(-x) - \pi$

(c) $h(x) = \tan(x + 2\pi) - 1$

(d) $j(x) = -\sin\left(2x + \frac{\pi}{4}\right)$

(e) $k(x) = -\cos\left(\frac{1}{4}x - 3\pi\right) + 2\pi$

(f) $l(x) = \tan\left(-\frac{2}{3}x + \frac{\pi}{2}\right)$

(g) $m(x) = -\sin(ex)$

(h) $n(x) = \cos\left(\pi^{-1}x - 1\right)$

(i) State the the transformations on a fundamental trigonometric function. Be sure to list them in the correct order.

(ii) State the period, phase shift, amplitude, and equilibrium position.

(iii) Sketch a graph of each transformed function. Your graph should include at least one full period.

(iv) State the domain and range of the transformed function.

(2) Write the equation of a cosine function that exhibits the following visual characteristics; period of 3, amplitude of 12, equilibrium position of -2, phase shift of π.

7.8. Inverse Trigonometric Functions

Our goal for this section is to develop inverse functions for sine, cosine, and tangent. To be invertible, a function must be bijective. Remember that all functions we will consider in $\mathbb{R}$ are onto. Working with sine, cosine, and tangent has provided us with some experience that none of these functions are 1-1. To illustrate that idea, consider the fact that

$$\sin\left(\frac{\pi}{4}\right) = \frac{\sqrt{2}}{2} \quad \text{and} \quad \sin\left(\frac{3\pi}{4}\right) = \frac{\sqrt{2}}{2}$$

This scenario is laid out visually below.

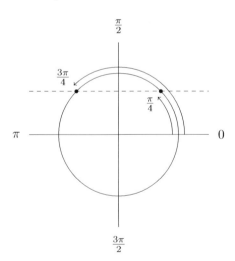

FIGURE 34. The endpoints for arcs $\frac{\pi}{4}$ and $\frac{3\pi}{4}$ share the same vertical value.

Further still, we know that any 2π revolution around the unit circle from the endpoints of arcs $\frac{\pi}{4}$ and $\frac{3\pi}{4}$ will have the same result. In other words

$$\sin\left(\frac{\pi}{4} + 2\pi k\right) = \frac{\sqrt{2}}{2} \quad \text{and} \quad \sin\left(\frac{3\pi}{4} + 2\pi k\right) = \frac{\sqrt{2}}{2}$$

where $k \in \mathbb{Z}$. Thus there are infinitely many domain values that pair with any single range value. We can solve this problem, and make the sine function 1-1, by restricting the domain.

For $f(x) = \sin x$, we choose a domain of $-\frac{\pi}{2} \le x \le \frac{\pi}{2}$. This restriction allows the sine function to receive inputs (arcs on the unit circle) only from quadrants one and four. The sine function is 1-1 on this restricted domain, and the range remains $-1 \le f(x) \le 1$. Figure 35 illustrates the domain restriction on both the unit circle, and the graph of the sine function.

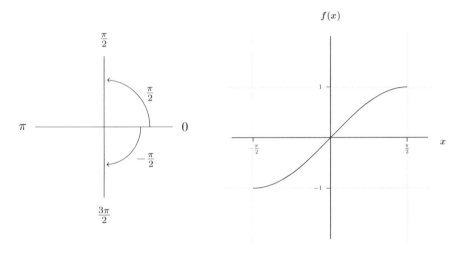

FIGURE 35. The restricted domain of $\left[-\frac{\pi}{2}, \frac{\pi}{2}\right]$ for the sine function.

Restricting the domain of sine does not change the definition of the sine function. The input to sine is still an arc on the unit circle. The domain restriction is a restriction on the arcs that can be used as input. The output of the sine function is still the vertical coordinate of the arc endpoint.

As this restriction makes the sine function 1-1 (and thus bijective), we can now consider an inverse. The inverse of sine is written

$$f(x) = \sin^{-1} x \qquad \text{or} \qquad f(x) = \arcsin x$$

The domain of inverse sine is

$$\{x \mid x \in \mathbb{R} \text{ and } -1 \le x \le 1\}$$

while the range is

$$\{f(x) \mid f(x) \in \mathbb{R} \text{ and } -\frac{\pi}{2} \le f(x) \le \frac{\pi}{2}\}$$

The inverse sine function maps in the opposite order of the sine function. Domain values of inverse sine are vertical coordinates for points on the unit circle (in quadrants one and four). Range values of inverse sine are arcs whose endpoints have the vertical coordinate given by the domain value. An example of this relationship is

$$\sin^{-1}\left(\frac{\sqrt{2}}{2}\right) = \frac{\pi}{4}$$

This pairing by inverse sine is a result of the fact that the vertical coordinate of the endpoint for an arc of length $\frac{\pi}{4}$ is $\frac{\sqrt{2}}{2}$. Or, to put it another way, because $\sin\left(\frac{\pi}{4}\right) = \frac{\sqrt{2}}{2}$.

As discussed earlier, there are infinitely many values for x such that $\sin(x) = \frac{\sqrt{2}}{2}$, but the domain restriction we placed on the sine function removes all but one, namely $x = \frac{\pi}{4}$. This fact is why there is exactly one answer to $\sin^{-1}\left(\frac{\sqrt{2}}{2}\right)$, and why that answer is $\frac{\pi}{4}$.

The domain restriction we placed on sine was required, as multiple arc endpoints share the same vertical value. It is also true that multiple arc endpoints share the same horizontal value. This fact can also be illustrated through an example. Consider that

$$\cos\left(\frac{\pi}{4}\right) = \frac{\sqrt{2}}{2} \quad \text{and} \quad \cos\left(\frac{7\pi}{4}\right) = \frac{\sqrt{2}}{2}$$

This relationship is shown below on the unit circle, and we know that any 2π revolution

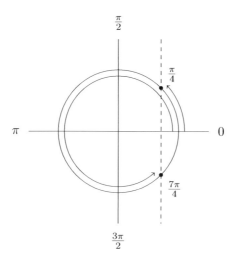

FIGURE 36. The endpoints for arc lengths $\frac{\pi}{4}$ and $\frac{7\pi}{4}$ share the same horizontal value.

from either of these endpoints will result in a new arc with the same horizontal value on its endpoint. In terms of the cosine function, this relationship would be written

$$\cos\left(\frac{\pi}{4} + 2\pi k\right) = \frac{\sqrt{2}}{2} \quad \text{and} \quad \cos\left(\frac{7\pi}{4} + 2\pi k\right) = \frac{\sqrt{2}}{2}$$

where $k \in \mathbb{Z}$. Thus, cosine is not 1-1, and we must restrict the domain to make the cosine function invertible. For $f(x) = \cos x$ we choose a domain restriction of $0 \leq x \leq \pi$. This restriction allows the cosine function to receive inputs (arcs on the unit circle) only from quadrants one and two. Note that the cosine function is now 1-1, and that the range remains $-1 \leq f(x) \leq 1$. Figure 37 illustrates the domain restriction on both the unit circle and the graph of the cosine function.

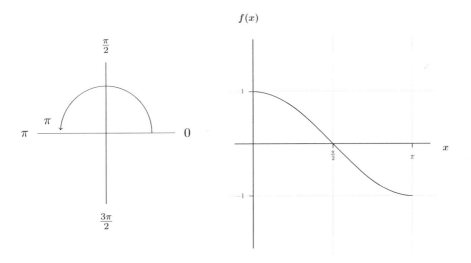

FIGURE 37. The restricted domain of $[0, \pi]$ for the cosine function.

The inverse of cosine is written

$$f(x) = \cos^{-1} x \quad \text{or} \quad f(x) = \arccos x$$

The domain of inverse cosine is

$$\{x \mid x \in \mathbb{R} \text{ and } -1 \leq x \leq 1\}$$

while the range is

$$\{f(x) \mid f(x) \in \mathbb{R} \text{ and } 0 \leq f(x) \leq \pi\}$$

Domain values of inverse cosine are horizontal coordinates for points on the unit circle (in quadrants one and two). Range values of inverse cosine are arcs whose endpoints have the horizontal coordinate given by the domain value. An example of this relationship would be

$$\cos^{-1}\left(\frac{\sqrt{2}}{2}\right) = \frac{\pi}{4}$$

which simply states that the arc length whose endpoint has a horizontal coordinate of $\frac{\sqrt{2}}{2}$ is $\frac{\pi}{4}$.

There are infinitely many values for x such that $\cos(x) = \frac{\sqrt{2}}{2}$, but the domain restriction we placed on the cosine function removes all but one, namely $x = \frac{\pi}{4}$. Thus, there is exactly one answer to $\cos^{-1}\left(\frac{\sqrt{2}}{2}\right)$ and that answer is $\frac{\pi}{4}$.

Many of the equations involving inverse trigonometric functions will require some form of technology to solve. As an example, suppose we wanted to find all solutions to the equation

$$\sin x = -0.6$$

Solving this equation is a direct application of the inverse sine function. We are solving for x, an arc on the unit circle, such that the vertical coordinate of the arc endpoint is -0.6. Figure 38 illustrates this scenario. Algebraically, the next step in this problem

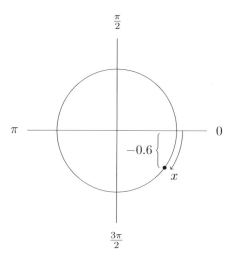

FIGURE 38. The endpoint of arc length x has a vertical coordinate of -0.6.

would to utilize the definition of inverse sine.

$$x = \sin^{-1}(-0.6)$$

The calculation of $\sin^{-1}(0.6)$ would be performed by some form of technology. Doing so tells us that

$$x \approx -0.643$$

or that an arc of length 0.643 radians, in the clockwise direction, will have an endpoint with a vertical value of -0.6. Note that there is another point in quadrant three that will have this same vertical value, but the range of inverse sine is restricted to quadrants one and two. The arc for the point in quadrant three must be located through our knowledge of the unit circle. Figure 39 illustrates the point in quadrant three, and the additional arc we must determine.

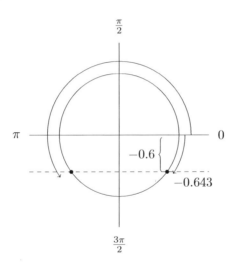

FIGURE 39. The additional solution in quadrant three.

A closer examination of Figure 39 suggests that we can separate the unknown arc into two pieces, one of which will have length π. An arc of length π ends at the point $(-1,0)$, and the distance from $(-1,0)$ to the endpoint of the unknown arc must be 0.643 (based on the symmetry of the unit circle). Thus, the unknown arc must be $\pi + 0.643$.

Any 2π revolution around the unit circle will arrive back at either of these respective endpoints. The full set of solutions is given by

$$x \approx -0.643 + 2\pi k \quad \text{or} \quad x \approx (\pi + 0.643) + 2\pi k$$

where $k \in \mathbb{Z}$.

Solving equations utilizing the inverse cosine function is analogous to our previous example involving inverse sine. We leave this as an exercise. Keep in mind that inverse cosine will provide an answer somewhere in quadrants one or two, and there will be an additional solution somewhere in quadrant three or four (exactly analogous to inverse sine providing a solution in quadrants one or four, and there being an additional solution in quadrants two or three).

Much like sine and cosine, the tangent function is not 1-1. Therefore, we will restrict the domain to alleviate this problem. The domain restriction for tangent will be $-\frac{\pi}{2} < x < \frac{\pi}{2}$. This restriction is shown visually in Figure 40. Note that even with a restricted domain, the range of the tangent function is still $\mathbb{R}$.

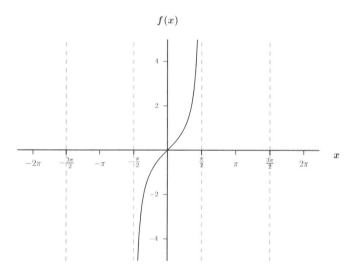

FIGURE 40. $f(x) = \tan x$ on domain $-\frac{\pi}{2} < x < \frac{\pi}{2}$

The inverse of tangent is given by

$$f(x) = \tan^{-1} x \quad \text{or} \quad f(x) = \arctan x$$

Rather than interpret inverse tangent through the unit circle, we will utilize the definition of an inverse function (from our previous work in function theory). It immediately follows that the domain and range of inverse tangent must be $\mathbb{R}$, and $\{f(x) \mid -\frac{\pi}{2} < f(x) < \frac{\pi}{2}\}$. The graph of $f(x) = \tan^{-1} x$ is shown in Figure 41.

Let us explore some examples of inverse tangent. Suppose we wish to evaluate

$$\tan^{-1}(-1)$$

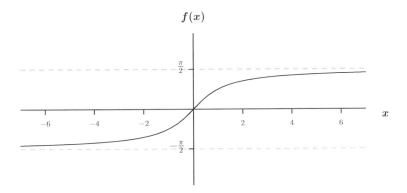

FIGURE 41. $f(x) = \tan^{-1} x$

What arc, θ, makes it true that $\tan \theta = -1$? Our work with the unit circle provides two answers,

$$\tan\left(\frac{3\pi}{4}\right) = -1 \quad \text{and} \quad \tan\left(\frac{7\pi}{4}\right) = -1$$

yet neither $\frac{3\pi}{4}$, nor $\frac{7\pi}{4}$ is in the range of inverse tangent. However, the arc $\theta = \frac{7\pi}{4}$, and the arc $\theta = -\frac{\pi}{4}$ both terminate at the same endpoint. Furthermore, $-\frac{\pi}{4}$ is in the range of inverse tangent. Thus, we have that $\tan^{-1}(-1) = -\frac{\pi}{4}$. Verify that this is a point on the graph of Figure 41.

As a final example, suppose we want to find all solutions to $\cot x = 3$. Let us begin by rewriting this equation in terms of tangent.

$$\cot x = 3$$

$$\frac{1}{\tan x} = 3$$

$$\frac{1}{3} = \tan x$$

We are trying to find an arc length, x, such that the tangent of that arc length is $\frac{1}{3}$. In terms of inverse tangent, we are trying to evaluate

$$\tan^{-1}\left(\frac{1}{3}\right) = x$$

We will rely on some sort of technology to calculate

$$\tan^{-1}\left(\frac{1}{3}\right) \approx 0.321$$

As tangent has a period of π, we have that all solutions for x are given by

$$x \approx 0.321 + k\pi$$

where $k \in \mathbb{Z}$.

Exercises for Section 7.8.

(1) To be invertible, sine, cosine, and tangent all required domain restrictions. Rather than using the restrictions that were given in this section, suppose we restrict the domain of sine to $[0, \pi]$, the domain of cosine to $\left[\frac{-\pi}{2}, \frac{\pi}{2}\right]$, and the domain of tangent to $[0, \pi]$. Would these new restrictions allow us to create inverse trigonometric functions? Why or why not?

(2) Evaluate each of the following using exact answers. Find all solutions on the interval $[0, 2\pi]$ and sketch your results on the unit circle.

(a) $\sin^{-1}\left(\frac{\sqrt{2}}{2}\right)$

(b) $\cos^{-1}\left(\frac{1}{2}\right)$

(c) $\arccos\left(\frac{\sqrt{3}}{2}\right)$

(d) $\arcsin\left(-\frac{\sqrt{3}}{2}\right)$

(e) $\sin^{-1}\left(\frac{1}{2}\right)$

(f) $\cos^{-1}\left(-\frac{\sqrt{2}}{2}\right)$

(3) This section developed the inverse tangent function directly from the general theory of an inverse function. Look back through Section 7.5, and revisit the unit circle interpretation of the tangent function. Use this interpretation to explain the pairings of inverse tangent, as well as the domain and range of inverse tangent.

(4) Find all solutions to the following equations. Use exact answers whenever possible.

(a) $\cos x = 0.75$ (b) $\sin x = 0$ (c) $\tan x = \sqrt{3}$

(5) Is it true that

$$\tan^{-1} x = \frac{\sin^{-1} x}{\cos^{-1} x}$$

Why or why not?

(6) Find the inverse of each of the following. State the range and the domain restriction you must utilize to make the function invertible.

(a) $f(x) = \cos(2x)$

(c) $h(x) = \tan(4x) + 1$

(b) $g(x) = 2\sin(x + 3\pi)$

(d) $j(x) = \sin x - \cos^2 x - \sin^2 x$

7.9. Compositions Involving Inverse Trigonometric Functions

Our previous work in function theory tells us that if f is an invertible function, then the compositions involving f and f^{-1} will have the relationships

$$f^{-1}(f(x)) = x \quad \text{on the invertible domain of } f$$

$$f(f^{-1}(x)) = x \quad \text{on the invertible domain of } f^{-1}$$

Let us explore these relationships with sine, cosine, and tangent.

Suppose we define a function f such that

$$f(x) = \sin\left(\sin^{-1} x\right)$$

The domain of this composition is the domain of inverse sine, $-1 \leq x \leq 1$. What is the range of this composition? We know that the range of inverse sine is the interval $\left[-\frac{\pi}{2}, \frac{\pi}{2}\right]$, and that this interval will become the input to the sine function. The sine of $\left[-\frac{\pi}{2}, \frac{\pi}{2}\right]$ is $[-1, 1]$. Thus, the range of the composition is $-1 \leq f(x) \leq 1$. The graph of this composition is shown in Figure 42. Note that $f(x) = \sin\left(\sin^{-1} x\right) = x$ on the entire domain for which this composition is defined.

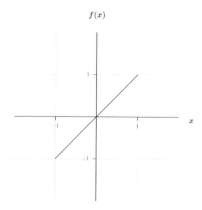

FIGURE 42. $f(x) = \sin\left(\sin^{-1} x\right)$

Now, let us consider a different composition

$$f(x) = \sin^{-1}(\sin x)$$

The domain of this composition is the domain of sine, $\mathbb{R}$. The output of the sine function is always on the interval $[-1, 1]$, so the output of sine is always a valid input to inverse sine. However, it is not the case that $f(x) = \sin^{-1}(\sin x) = x$ on a domain of $\mathbb{R}$. In fact, $f(x) = \sin^{-1}(\sin x) = x$ on a domain of $-\frac{\pi}{2} \leq x \leq \frac{\pi}{2}$. Why? The first step in our development of the inverse sine function was to restrict the domain of the sine function. The invertible domain of sine is $-\frac{\pi}{2} \leq x \leq \frac{\pi}{2}$, thus that is the domain on which the relationship

$$\sin^{-1}(\sin x) = x$$

holds true. The graph of $f(x) = \sin^{-1}(\sin x)$ is shown in Figure 43. Note that $f(x) = \sin^{-1}(\sin x) = x$ on a domain of $-\frac{\pi}{2} \leq x \leq \frac{\pi}{2}$, not the domain of $\mathbb{R}$ for which this composition is defined.

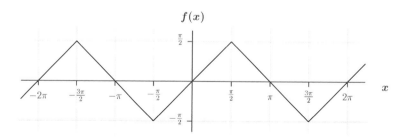

FIGURE 43. $f(x) = \sin^{-1}(\sin x)$

7.9. COMPOSITIONS INVOLVING INVERSE TRIGONOMETRIC FUNCTIONS

Compositions of cosine with inverse cosine, and tangent with inverse tangent are analogous to those of sine. The composition

$$f(x) = \cos\left(\cos^{-1} x\right)$$

has domain $-1 \leq x \leq 1$, and $f(x) = \cos\left(\cos^{-1} x\right) = x$ on the entire domain. The graph of this composition is given in Figure 44.

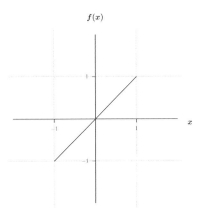

FIGURE 44. $f(x) = \cos\left(\cos^{-1} x\right)$

Changing the order of the composition

$$f(x) = \cos^{-1}(\cos x)$$

gives a function with domain $\mathbb{R}$. Now, however, $f(x) = \cos^{-1}(\cos x) = x$ only on a domain of $0 \leq x \leq \pi$, the invertible domain of the cosine function. Figure 45 illustrates this relationship.

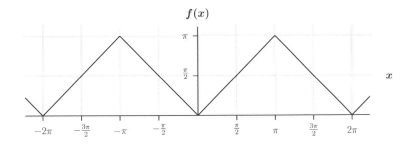

FIGURE 45. $f(x) = \cos^{-1}(\cos x)$

For the function

$$f(x) = \tan\left(\tan^{-1} x\right)$$

the domain is $\mathbb{R}$, and $f(x) = \tan\left(\tan^{-1} x\right) = x$ on the entire domain (Figure 46).

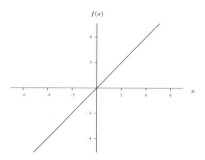

FIGURE 46. $f(x) = \tan\left(\tan^{-1} x\right)$

Changing the composition order to

$$f(x) = \tan^{-1}(\tan x)$$

results in a domain of $\{x \mid x \neq \frac{\pi}{2} + k\pi \text{ where } k \in \mathbb{Z}\}$. Note that $f(x) = \tan^{-1}(\tan x) = x$ only on $-\frac{\pi}{2} < x < \frac{\pi}{2}$, the invertible domain of the tangent function (Figure 47).

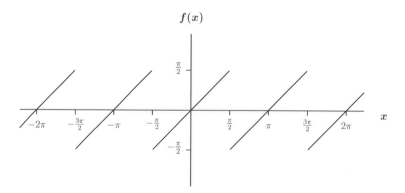

FIGURE 47. $f(x) = \tan^{-1}(\tan x)$

Let us use our knowledge of inverse functions and composition to explore some examples. Suppose we wish to find the exact value of

$$\sin\left(\cos^{-1}\left(\frac{4}{5}\right)\right)$$

7.9. COMPOSITIONS INVOLVING INVERSE TRIGONOMETRIC FUNCTIONS

As always, we start by considering the innermost parentheses. The inverse cosine of $\frac{4}{5}$ is an arc, such that the horizontal coordinate of the endpoint is $\frac{4}{5}$. If we refer to this arc as θ, we can write the relationship

$$\theta = \cos^{-1}\left(\frac{4}{5}\right)$$

from which it immediately follows that

$$\cos\theta = \frac{4}{5}$$

We must be cautious before continuing. Theta was defined as the inverse cosine of $\frac{4}{5}$. The range of inverse cosine is limited to quadrants one and two, the interval $[0, \pi]$. On a single revolution of the unit circle, $[0, 2\pi]$, there are two arcs such that the horizontal coordinate of the endpoint is $\frac{4}{5}$. One of these arcs terminates in quadrant one, the other will terminate in quadrant four. We cannot consider the answer that terminates in quadrant four, as it is outside the range of inverse cosine. In other words, it must be the case that θ terminates in quadrant one.

Substitution of $\theta = \cos^{-1}\left(\frac{4}{5}\right)$ into the original expression gives

$$\sin\left(\cos^{-1}\left(\frac{4}{5}\right)\right) = \sin\theta$$

As we know $\cos\theta = \frac{4}{5}$, we can use the relationship $\sin^2\theta + \cos^2\theta = 1$ to solve for $\sin\theta$.

$$\sin^2\theta + \left(\frac{4}{5}\right)^2 = 1$$

$$\sin^2\theta = 1 - \left(\frac{4}{5}\right)^2$$

$$\sin\theta = \pm\sqrt{1 - \left(\frac{4}{5}\right)^2}$$

$$\sin\theta = \pm\sqrt{1 - \frac{16}{25}}$$

$$\sin\theta = \pm\sqrt{\frac{9}{25}}$$

$$\sin\theta = \pm\frac{3}{5}$$

We previously established θ terminates in quadrant one, so the sine of theta must be positive. Thus, $\sin\theta = \frac{3}{5}$. In summary

$$\sin\left(\cos^{-1}\left(\frac{4}{5}\right)\right) = \frac{3}{5}$$

As another example, suppose we wish to express

$$\tan\left(\sin^{-1}(x-1)\right)$$

in terms of x without utilizing any trigonometric functions. As we know the output of the inverse sine function is an arc, we can let

$$\theta = \sin^{-1}(x-1)$$

Theta is an arc, such that the vertical coordinate of the endpoint is $x-1$. We know θ must terminate somewhere in the interval covered by quadrants one and four (the range of the inverse sine function), but as we do not know if $x-1 > 0$, $x-1 < 0$, or if $x-1 = 0$, we do not know if θ terminates in quadrant one, four, or on the axis. This scenario is shown graphically in Figure 48.

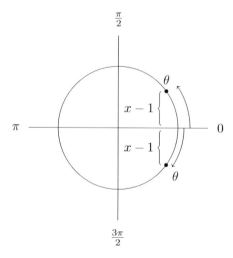

FIGURE 48. Possible termination quadrants for $\theta = \sin^{-1}(x-1)$.

We can substitute θ into our original expression to give

$$\tan\left[\sin^{-1}(x-1)\right] = \tan\theta = \frac{\sin\theta}{\cos\theta}$$

Despite not knowing the termination location of θ, we still have the relationship

$$\sin \theta = x - 1$$

and we can solve for $\cos \theta$ using the identity $\sin^2 \theta + \cos^2 \theta = 1$.

$$(x-1)^2 + \cos^2 \theta = 1$$

$$x^2 - 2x + 1 + \cos^2 \theta = 1$$

$$\cos^2 \theta = -x^2 + 2x$$

$$\cos \theta = \pm\sqrt{-x^2 + 2x}$$

As illustrated in Figure 48, the horizontal coordinate of the endpoint for θ is positive in quadrants one and four. This means we need to select the positive solution for $\cos \theta$. Substituting the expressions for $\sin \theta$ and $\cos \theta$ into our original problem gives

$$\tan\left[\sin^{-1}(x-1)\right] = \tan \theta = \frac{\sin \theta}{\cos \theta} = \frac{x-1}{\sqrt{-x^2 + 2x}}$$

and we have expressed our original problem in terms of x without the use of any trigonometric functions.

For a final example, let us solve the equation

$$\sin^{-1}(3x) = \cos^{-1}(x+1)$$

The output of both inverse sine and inverse cosine is an arc. In this example, we have an inverse sine set equal to an inverse cosine. This means that the arc corresponding to both of these inverse trigonometric functions must be the same. In other words

$$\theta = \sin^{-1}(3x) \quad \text{and} \quad \theta = \cos^{-1}(x+1)$$

from which we can obtain the relationships

$$\sin \theta = 3x \quad \text{and} \quad \cos \theta = x+1$$

Next, we need to determine where theta terminates. As we know θ is the output of both an inverse sine and an inverse cosine, it must be the case that θ terminates in a region that is common to the range of inverse sine and inverse cosine. The range of inverse sine is $\left[-\frac{\pi}{2}, \frac{\pi}{2}\right]$, and the range of inverse cosine is $[0, \pi]$. The overlap of these two ranges is $\left[0, \frac{\pi}{2}\right]$, so θ must terminate somewhere in this interval.

Once again, we can utilize the identity $\sin^2 \theta + \cos^2 \theta = 1$ to solve for x.

$$(3x)^2 + (x+1)^2 = 1$$

$$9x^2 + x^2 + 2x + 1 = 1$$

$$10x^2 + 2x = 0$$

$$2x(5x+1) = 0$$

from which we conclude $x = 0$ or $x = -\frac{1}{5}$. The solution $x = -\frac{1}{5}$ would result in

$$\sin \theta = 3x = 3\left(-\frac{1}{5}\right) = -\frac{3}{5}$$

which would correspond to an arc outside of the interval $[0, \pi]$. As we know that θ must terminate on this interval, we can disregard the solution $x = -\frac{1}{5}$, and the only solution to the equation is $x = 0$.

Exercises for Section 7.9.

(1) Explain the shape of the graphs

$$f(x) = \sin^{-1}(\sin x) \quad \text{and} \quad g(x) = \cos^{-1}(\cos x)$$

on the interval $[0, 2\pi]$

(2) Evaluate each of the following. Use exact answers whenever possible.

(a) $\sin\left(\cos^{-1}\left(\frac{1}{3}\right)\right)$

(c) $\cos^{-1}\left(\cos\left(\frac{\sqrt{2}}{2}\right)\right)$

(b) $\cos\left(\sin^{-1}\left(-\frac{2}{5}\right)\right)$

(d) $\sin\left(\sin^{-1}(2)\right)$

(3) Determine the domain and range of the functions

$$f(x) = \sin\left(\cos^{-1} x\right) \quad \text{and} \quad g(x) = \cos\left(\sin^{-1} x\right)$$

(4) On a unit circle, illustrate the following relationship
$$\cos^{-1}(\sin 2.5) \approx 0.929$$

(5) Rewrite the following expressions without trigonometric functions. Consider what values x can take in both the trigonometric and algebraic versions of these expressions. Do they agree?

(a) $\sin\left(\cos^{-1}(2x+1)\right)$

(b) $\tan\left(\cos^{-1}(x+4)\right)$

(6) Solve the following equations.

(a) $\sin^{-1}(x-3) = \cos^{-1}(x-2)$

(b) $\cos\left(\sin^{-1} x + \cos^{-1}(2x-1)\right) = -\dfrac{1}{2}$

Hint: The function
$$f(x) = 13x^4 - 28x^3 + \frac{41}{2}x^2 - 6x + \frac{9}{16}$$
has three zeros. One of the zeros is less than one half, one of the zeros is one half, and the third zero is $x \approx 0.976$.

7.10. Extending to Non-Unit Circles

This section will revisit the definitions of sine and cosine for right triangles and discuss how we can use them to extend our working knowledge of sine and cosine to non-unit circles. Let us begin by considering the radian version of the 30-60-90 special right triangle.

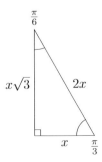

Any value we use for x changes all legs of the triangle by the same factor of x. This fact is why we can use right triangle trigonometry to say that

$$\cos\frac{\pi}{3} = \frac{x}{2x} = \frac{1}{2}$$

regardless of the triangle we are making reference to. This idea can be extended to all right triangles (not just the special ones).

Consider any right triangle. If c represents the length of the hypotenuse, and a and b the lengths of the remaining sides, the Pythagorean theorem tells us

$$a^2 + b^2 = c^2$$

Solving for the length of c would give

$$c = \sqrt{a^2 + b^2}$$

Now, suppose we scale c by a factor of n, where n is any positive real number.

$$nc = n\sqrt{a^2 + b^2}$$
$$= \sqrt{n^2(a^2 + b^2)}$$
$$= \sqrt{n^2 a^2 + n^2 b^2}$$
$$= \sqrt{(na)^2 + (nb)^2}$$

The end result is that the lengths of both other legs are scaled by a factor of n as well. We can use this fact to evaluate arc endpoint coordinates on non-unit circles.

Suppose we have a circle of radius two. Start at the point $(2, 0)$, and moving in the counter clockwise direction along an arc of length 4.5. What are the coordinates of the endpoint? Let us begin by drawing a picture, Figure 49.

The coordinates of the arc endpoint are not $(\cos 4.5, \sin 4.5)$. Those coordinates would correspond to the endpoint of an arc of length 4.5 on the unit circle, not a circle of radius two. To calculate the corresponding arc length on the unit circle, we need to convert 4.5 into radians.

$$\theta = \frac{\text{arc length}}{r} = \frac{4.5}{2} = 2.25$$

Adding the unit circle with an arc of 2.25 to our picture is shown in Figure 50. All that remains is to fill in the corresponding right triangles.

7.10. EXTENDING TO NON-UNIT CIRCLES

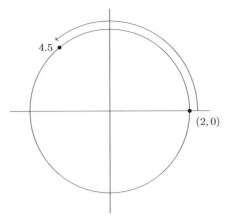

FIGURE 49. An arc of length 4.5 on a circle with radius 2.

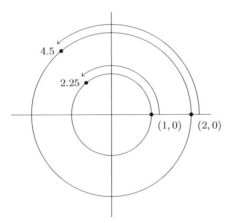

FIGURE 50. Adding the unit circle to Figure 49.

The legs of the right triangle inscribed in the circle of radius two are exactly double the legs inscribed the unit circle. On the unit circle, the coordinates of the endpoint for the arc of 2.25 are

$$(\cos 2.25, \sin 2.25)$$

Thus, on the circle of radius two, the coordinates of the endpoint for the arc of 4.5 must be

$$(2(\cos 2.25), 2(\sin 2.25))$$

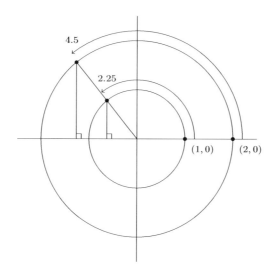

FIGURE 51. Adding right triangles to Figure 50.

All of the methods in the previous example are not dependent on a radius of two. In fact, this approach can generalized for any circle. When solving for the endpoint of any arc on any circle, convert the arc length into radians, calculate the corresponding endpoint on the unit circle, and then scale the endpoint based on the radius of the original circle.

Exercises for Section 7.10.

(1) Determine the endpoints of the following arcs.

    (a) An arc of π on a circle of radius 3.

    (b) An arc of -5 on a circle of radius 6

(2) Suppose you have an arc of length A on a circle of radius r, centered at the origin. What are the coordinates for the endpoint of the arc?

(3) Assume that the Earth is a perfect sphere (in reality, it bulges along the equator), with radius $r = 6,378$ km. Suppose you are in an airplane traveling 900 km/hour, and the plane is flying along the equator. At 12 pm, the plane is in location A. After five hours, the plane is in location B.

    Suppose also that you have access to a fantastical drilling machine, capable of tunneling through the Earth. How much shorter is the direct trip, tunneling through the Earth, between A and B? Disregard the height of the plane, and the fact that you will most certainly perish in this tunneling adventure.

(4) Reconsider the previous problem, but with the following changes. Now, the plane is flying at an altitude of 11 km, and it is no longer flying directly above the equator. Discuss how these changes alter the previous problem, and compare the distance the plane travels with the distance the tunneling machine travels.

CHAPTER 8

Trigonometric Equations and Oblique Triangles

The first few sections of this chapter are devoted to solving equations that involve trigonometric functions. An example of a simple trigonometric equation is

$$\cos x = 0.7$$

which is solved utilizing the inverse cosine function, and some knowledge of the unit circle. Our work in this chapter will expand these types of equations, develop some useful trigonometric identities, and ultimately allow us to solve non-right triangles.

8.1. Trigonometric Equalities and Inequalities

We begin this section by considering a slightly more complicated version of the equation above.

$$\cos\left(\frac{3x-5}{2}\right) = 0.7$$

Though the argument of this cosine equation is no longer a simple x, the fundamental relationship of the equation is unchanged. Informally, both equations say "the horizontal coordinate of the endpoint for some unknown arc is 0.7". As such, we approach solving both equations in the same fashion, through use of the inverse cosine function.

$$\frac{3x-5}{2} = \cos^{-1}(0.7)$$

Our previous experience with the inverse cosine function tells us

$$\frac{3x-5}{2} \approx \pm 0.795 + 2\pi k \qquad \text{where } k \in \mathbb{Z}$$

At this point, solving for x is straightforward algebra.

$$3x - 5 \approx \pm 1.59 + 4\pi k$$

$$3x \approx 5 \pm 1.59 + 4\pi k$$

$$x \approx \frac{5}{3} \pm \frac{1.59}{3} + \frac{4\pi}{3}k$$

Applying the plus or minus leaves us with

$$x \approx \frac{6.59}{3} + \frac{4\pi}{3}k \quad \text{or} \quad x \approx \frac{3.41}{3} + \frac{4\pi}{3}k$$

Now that we have found all solutions to our equation, let us take a moment and revisit the argument to the cosine function.

$$\cos\left(\frac{3x-5}{2}\right) = \cos\left(\frac{3}{2}x - \frac{5}{2}\right)$$

In terms of transformations on a standard cosine function, this argument would correspond to a right shift by $\frac{5}{2}$, and a horizontal compression by a factor of $\frac{2}{3}$. A compression of $\frac{2}{3}$ on the period of the standard cosine function would result in a period of

$$\frac{2}{3}(2\pi) = \frac{4\pi}{3}$$

which is exactly the period we see in our answers.

Moving on, let us find all solutions between 0 and 2π to the equation

$$\cos(2x) = \frac{3}{5}$$

The initial step is once again the use of an inverse trigonometric function.

$$2x = \cos^{-1}\left(\frac{3}{5}\right)$$

$$2x \approx \pm 0.927 + 2\pi k$$

$$x \approx \pm 0.463 + \pi k$$

and we arrive at the conclusion that all solutions to the equation are given by

$$x \approx 0.463 + \pi k \quad \text{or} \quad x \approx -0.463 + \pi k$$

where $k \in \mathbb{Z}$. As we are solving for solutions between 0 and 2π, we need only consider values for k that provide answers on that interval.

$$x \approx 0.463 + \pi(0) \approx 0.436 \qquad x \approx -0.463 + \pi(1) \approx 2.677$$

$$x \approx 0.463 + \pi(1) \approx 3.605 \qquad x \approx -0.463 + \pi(2) \approx 5.819$$

These four solutions represent all answers on the interval 0 to 2π. Note that the values for k do not need to be the same between the positive and negative answers.

Now, let us solve a trigonometric inequality.

$$\sin x > 0.7$$

Figure 1 illustrates this scenario. All arcs that terminate above the horizontal line in

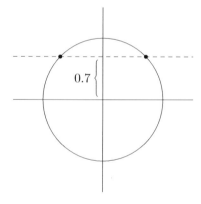

FIGURE 1. Arcs where the vertical coordinate of the endpoint is greater that 0.7 would all terminate above the horizontal line.

Figure 1 will have an endpoint such that the vertical coordinate of the endpoint is greater than 0.7. We begin by solving for the arcs whose endpoints have a vertical coordinate of 0.7.

$$\sin x = 0.7$$

$$x = \sin^{-1}(0.7)$$

from which we arrive at the solutions

$$x \approx 0.775 \quad \text{or} \quad x \approx \pi - 0.775$$

Both arcs are shown together in Figure 2.

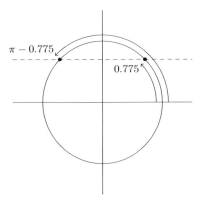

FIGURE 2. Arcs between these two will have an endpoint such that the vertical coordinate of the endpoint is greater than 0.7.

The solution to the inequality consists of all arcs between these two solutions. Writing this relationship out mathematically leaves us with

$$0.775 < x < \pi - 0.775$$

This inequality consists of all arcs on the first revolution of the unit circle such that the vertical coordinate of the endpoint is greater than 0.7. All additional revolutions of the unit circle can be included in this answer by adding $2\pi k$, where $k \in \mathbb{Z}$.

$$0.775 + 2\pi k < x < (\pi - 0.775) + 2\pi k$$

Our next example illustrates use of the zero factor property. The equation

$$\sin^2 x - 3\sin x = 0$$

can be factored into

$$(\sin x)(\sin x - 3) = 0$$

from which the zero factor property immediately tells us that

$$\sin x = 0 \quad \text{or} \quad \sin x - 3 = 0$$

There is no solution to the equation $\sin x - 3 = 0$. To understand why, consider what happens if we move the three to the right side of the equation.

$$\sin x = 3$$

This equation states that there is an arc on the unit circle such that the vertical coordinate of the arc endpoint is equal to 3. This statement is false. The maximum vertical value for any point on the unit circle is 1.

Solving $\sin x = 0$ is identifying all arcs such that the vertical coordinate of the arc endpoint is 0. Any arc whose length is a multiple of π will satisfy this condition. Thus, the answers to $\sin x = 0$ are given by

$$x = \pi k$$

where $k \in \mathbb{Z}$.

As a final example, consider the equation

$$\cos^2 x + \cos x - 1 = 0$$

Solving this equation will be our first detailed experience of *variable substitution*, which allows us to use a temporary variable to take the place of a more complicated expression. In the context of this problem, we will let

$$u = \cos x$$

We have utilized the variable u as it is an extremely common choice for variable substitution in calculus. Substituting u into our original equation gives

$$u^2 + u - 1 = 0$$

which can be solved with the quadratic formula

$$u = \frac{-1 \pm \sqrt{5}}{2}$$

At this point, we have solved for u. As the original problem was in terms of x, we need to substitute our value for u (which is in terms of x) into these answers.

$$\cos x = \frac{-1 + \sqrt{5}}{2} \quad \text{or} \quad \cos x = \frac{-1 - \sqrt{5}}{2}$$

As was the case in the previous example, there are no solutions to one of these equations. As the quantity

$$\frac{-1 - \sqrt{5}}{2} < -1$$

there is no way it could represent the horizontal coordinate of a point on the unit circle. Utilizing the inverse cosine function on the remaining equation will leave us with all possible solutions for x

$$x = \cos^{-1}\left(\frac{-1+\sqrt{5}}{2}\right) \approx \pm 0.904 + 2\pi k$$

where $k \in \mathbb{Z}$.

Exercises for Section 8.1.

(1) Solve the following equations. Identify all answers and explicitly state the answers on the interval $[0, 2\pi]$. Find exact solutions when possible.

(a) $4\cos\left(\dfrac{3x-1}{2}\right) = 1$

(b) $\sin\left(\dfrac{2x}{5}\right) - 1 = -\dfrac{3}{2}$

(c) $\dfrac{\tan(\pi x)}{3} = 4$

(d) $2\cos\left(3x - \dfrac{\pi}{2}\right) = 2$

(2) Find all solutions to the following inequalities.

(a) $2\cos\left(x - \dfrac{\pi}{4}\right) \geq 1$

(b) $\sin x > \cos x$

(c) $\sin(x - \pi) < 0.5$

(d) $\cos^2(x+1) > \dfrac{3}{4}$

(3) Find all solutions to the following equations. Use exact answers whenever possible.

(a) $2\sin^2 x = \sin x$

(b) $4\cos^2(3x) - \cos(3x) = 3$

(c) $e^{\cos x} = 2$

(d) $\sin x = \tan x$

(e) $3\tan^2 x + 4\tan x = 1$

(f) $\cos x - 3\cos^2 x = 2$

8.2. Trigonometric Identities

Our work thus far has provided experience with several of the fundamental identities of trigonometry.

$$\sin(\alpha + \beta) = \sin\alpha\cos\beta + \cos\alpha\sin\beta \qquad \sin(-\theta) = -\sin\theta$$

$$\sin(\alpha - \beta) = \sin\alpha\cos\beta - \cos\alpha\sin\beta \qquad \cos(-\theta) = \cos\theta$$

$$\cos(\alpha + \beta) = \cos\alpha\cos\beta - \sin\alpha\sin\beta \qquad \sin^2\theta + \cos^2\theta = 1$$

$$\cos(\alpha - \beta) = \cos\alpha\cos\beta + \sin\alpha\sin\beta$$

The purpose of this section is to explore the process of discovering trigonometric identities, and to practice their use. There are many different identities in trigonometry. While it is certainly possible to sit down and memorize a long list of these identities, a better approach is to become comfortable and confident working with definitions and a few fundamental identities, and then use these tools to derive trigonometric relationships as you need them.

To illustrate this process, consider the fundamental identity $\sin^2\theta + \cos^2\theta = 1$. If we divide both sides of this identity by $\cos^2\theta$

$$\frac{\sin^2\theta}{\cos^2\theta} + \frac{\cos^2\theta}{\cos^2\theta} = \frac{1}{\cos^2\theta}$$

$$\tan^2\theta + 1 = \sec^2\theta$$

we arrive at a new relationship. While this result can be useful in certain situations, is it worth committing this identity to memory? Potentially not. Knowing the fundamental identity we started with, and the definitions of tangent and secant is enough to quickly derive this identity if you should need it.

The *double angle* identities are extremely common and useful trigonometric relationships. Both follow quickly from the addition identities, and both are worth committing to memory.

$$\sin(2\theta) = \sin(\theta + \theta) \qquad\qquad \cos(2\theta) = \cos(\theta + \theta)$$

$$= \sin\theta\cos\theta + \sin\theta\cos\theta \qquad = \cos\theta\cos\theta - \sin\theta\sin\theta$$

$$= 2\sin\theta\cos\theta \qquad\qquad = \cos^2\theta - \sin^2\theta$$

Let us get some practice with the double angle identities. Suppose we know that $\cos\theta = \frac{2}{5}$, and we want to find the exact value of $\cos(2\theta)$.

$$\cos(2\theta) = \cos^2\theta - \sin^2\theta$$

$$= \left(\frac{2}{5}\right)^2 - \sin^2\theta$$

$$= \frac{4}{25} - \sin^2\theta$$

If we can determine $\sin^2\theta$, we can provide an exact answer. Another identity, $\sin^2\theta + \cos^2\theta = 1$, will allow us to calculate $\sin^2\theta$.

$$\sin^2\theta = 1 - \cos^2\theta$$

$$= 1 - \left(\frac{2}{5}\right)^2$$

$$= \frac{25}{25} - \frac{4}{25}$$

$$= \frac{21}{25}$$

Plugging this in to the original expression

$$\cos(2\theta) = \frac{4}{25} - \sin^2\theta = \frac{4}{25} - \frac{21}{25} = -\frac{17}{25}$$

Alternatively, we could approach the previous problem slightly differently. Utilizing the same identities, we can make the following algebraic substitutions.

$$\cos(2\theta) = \cos^2\theta - \sin^2\theta$$

$$= \cos^2\theta - (1 - \cos^2\theta)$$

$$= 2\cos^2\theta - 1$$

Using the known value of $\cos\theta = \frac{2}{5}$

$$\cos(2\theta) = 2\left(\frac{2}{5}\right)^2 - 1 = \frac{8}{25} - \frac{25}{25} = -\frac{17}{25}$$

There will often be multiple approaches to solving trigonometric equations. Regardless of the method you choose, if your steps are mathematically correct, you will always arrive at the correct answer. That having been said, there will often be an approach that is simpler than the rest.

Suppose we wish to solve

$$3\cos(2x) + \cos(-2x) + 1 = 0.4$$

Consider the different approaches we could take.

(1) Use the double angle identity for cosine.
(2) Use the fact that cosine is an even function.

While both of these are algebraically correct, option two will result in an easier path to the solution. If we apply the fact that cosine is an even function, we have $\cos(-2x) = \cos(2x)$.

$$3\cos(2x) + \cos(-2x) + 1 = 0.4$$

$$3\cos(2x) + \cos(2x) + 1 = 0.4$$

$$4\cos(2x) = -0.6$$

$$\cos(2x) = -0.15$$

$$2x = \cos^{-1}(-0.15)$$

$$2x \approx \pm 1.721 + 2\pi k$$

$$x \approx \pm 0.86 + \pi k$$

where $k \in \mathbb{Z}$.

We will continue to practice variable substitution throughout our work in trigonometry because it plays a large role in calculus. So now, let us solve the equation

$$2\sin x + \cos^2 x = 0$$

We begin by utilizing the fact that $\cos^2 x = 1 - \sin^2 x$. Rewriting the cosine in terms of sine leaves us an equation with only one trigonometric function, sine.

$$2 \sin x + \cos^2 x = 0$$

$$2 \sin x + 1 - \sin^2 x = 0$$

$$\sin^2 x - 2 \sin x - 1 = 0$$

At this point, a variable substitution of $u = \sin x$ will allow us to treat our equation like a quadratic.

$$u^2 - 2u - 1 = 0$$

The quadratic formula will allow us to solve for u.

$$u = \frac{2 \pm \sqrt{8}}{2} = 1 \pm \sqrt{2}$$

Now, as $u = \sin x$.

$$\sin x = 1 \pm \sqrt{2}$$

Separating the plus and minus solutions leaves us with

$$\sin x = 1 + \sqrt{2} \quad \text{or} \quad \sin x = 1 - \sqrt{2}$$

Any solution for x in the equation $\sin x = 1 + \sqrt{2}$ would correspond to an arc on the unit circle, such that the vertical coordinate of the arc endpoint was equal to $1 + \sqrt{2} \approx 2.4142$. As such, there are no solutions to that equation.

There are solutions to

$$\sin x = 1 - \sqrt{2}$$

$$x = \sin^{-1}\left(1 - \sqrt{2}\right)$$

From which it follows that

$$x \approx -0.427 + 2\pi k \quad \text{or} \quad x \approx (\pi + 0.427) + 2\pi k$$

where $k \in \mathbb{Z}$.

The use of a a single trigonometric identity will not always allow the immediate solution of an equation. Some problems will require the manipulation of multiple identities. To that end, let us solve

$$\sin(x) 4\cos^2(x) - \sin(x)\cos(2x) + 3\sin(x)\cos(x) = 0$$

As always, begin by considering potential options. Note that each term contains a factor of $\sin x$.

$$(\sin x)\left[4\cos^2 x - \cos(2x) + 3\cos x\right] = 0$$

If we can rewrite $\cos(2x)$ in terms of $\cos(x)$, then the factor enclosed in brackets would algebraically resemble a quadratic. Use of the double angle identity for cosine tells us

$$\cos(2x) = \cos^2 x - \sin^2 x$$

But, utilizing this identity will introduce a $\sin^2 x$ term, and we would very much like to keep things in terms of cosine. It is at this point that we must utilize a second identity. Specifically, $\sin^2 x + \cos^2 x = 1$. We can use this identity to write sine squared as

$$\sin^2 x = 1 - \cos^2 x$$

and then make this substitution into cosine's double angle identity

$$\cos(2x) = \cos^2 x - \sin^2 x$$
$$= \cos^2 x - (1 - \cos^2 x)$$
$$= 2\cos^2 x - 1$$

Utilizing this result in our equation

$$(\sin x)\left[4\cos^2 x - \cos(2x) + 3\cos x\right] = 0$$
$$(\sin x)\left[4\cos^2 x - (2\cos^2 x - 1) + 3\cos x\right] = 0$$
$$(\sin x)\left[2\cos^2 x + 3\cos x + 1\right] = 0$$
$$(\sin x)(2\cos x + 1)(\cos x + 1) = 0$$

The last step factored $2\cos^2 x + 3\cos x + 1$ exactly as you would factor the quadratic $2x^2 + 3x + 1$. We can now utilize the zero factor property and conclude

$$\sin x = 0 \quad \text{or} \quad 2\cos x + 1 = 0 \quad \text{or} \quad \cos x + 1 = 0$$

Solving each of these in turn would leave us with

$$x = \pi k \quad \text{or} \quad x = \pm \frac{2\pi}{3} + 2\pi k \quad \text{or} \quad x = \pi + 2\pi k$$

Note that the solutions of $\sin x = 0$ are given as $x = \pi k$ rather than $x = 0 + 2\pi k$ or $x = \pi + 2\pi k$. Both versions of the answer are correct. Additionally, the solutions of $x = \pi + 2\pi k$ are included in the solutions of $x = \pi k$. In summary, all solutions are given by

$$x = \pi k \quad \text{or} \quad x = \pm \frac{2\pi}{3} + 2\pi k$$

Exercises for Section 8.2.

(1) Prove these relationships

$$\sin\left(\frac{\pi}{2} - \theta\right) = \cos\theta \qquad \cos\left(\frac{\pi}{2} - \theta\right) = \sin\theta$$

(2) The double angle identity for tangent is given by

$$\tan(2\theta) = \frac{2\tan\theta}{1 - \tan^2\theta}$$

Prove this result.

(3) The triple angle identities for sine and cosine are given by

$$\sin(3\theta) = 3\sin\theta - 4\sin^3\theta \qquad \cos(3\theta) = 4\cos^3\theta - 3\cos\theta$$

Prove these results.

(4) Find all solutions to the following equations. Use exact answers whenever possible.

(a) $\cos x - \cos(2x) = 0$

(b) $\cos^4 x - \sin^4 x = 1$

(c) $\sin(x+\pi) - \cos(x+\pi) = 0$

(d) $\sin(2x) - \cos(2x) = -1$

(e) $\cos(4x)\sin(x) + \sin(x)\cos(4x) = 2$

Hint: $f(x) = 8x^5 - 8x^3 + x - 1$ has one zero, $x = 1$.

8.3. The Law of Cosines

The trigonometric relationships we have established for triangles apply exclusively to right triangles. Consider the oblique triangle $\triangle ABC$ shown below.

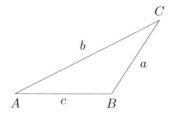

It is no longer true that

$$\sin A = \frac{\text{Length of } \overline{BC}}{\text{Length of } \overline{AC}} = \frac{a}{b}$$

as the right triangle definition of sine requires a right triangle. Similarly, the relationships given by cosine, tangent, and the Pythagorean Theorem are no longer applicable. Our goal in this section is to discover relationships between angles and side length in oblique triangles. To that end, suppose we draw a line from $\overline{AC}$ to $\angle B$ in the following fashion.

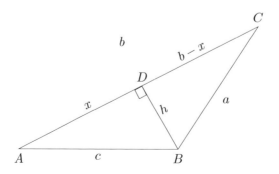

8.3. THE LAW OF COSINES

We now have two right triangles, and access to all of our right triangle trigonometry. In right triangle $\triangle BCD$ we have the relationship

(8.1) $$(b - x)^2 + h^2 = a^2$$

while in right triangle $\triangle ABD$ we have

(8.2) $$x^2 + h^2 = c^2$$

We can rewrite equation 8.2 as $h^2 = c^2 - x^2$ and substitute this relationship into equation 8.1.

$$(b - x)^2 + h^2 = a^2$$
$$(b - x)^2 + c^2 - x^2 = a^2$$
$$b^2 - 2bx + x^2 + c^2 - x^2 = a^2$$

(8.3) $$b^2 + c^2 - 2bx = a^2$$

Remember, our goal is to determine a relationship between the sides and angles of an oblique triangle. If we can rewrite the x in equation 8.3 as some combination of $a, b, c, \angle A, \angle B$, or $\angle C$, we will have accomplished our task.

To that end, note that in right triangle $\triangle ABD$

$$\cos A = \frac{x}{c}$$

solving this relationship for x gives

$$x = c \cos A$$

and making this substitution into equation 8.3 leaves us with

$$a^2 = b^2 + c^2 - 2bc \cos A$$

This relationship is the first of three parts that together are known as the *Law of Cosines*. Given any triangle with sides a, b, c and angles A, B, C, the Law of Cosines states

(1) $a^2 = b^2 + c^2 - 2bc \cos A$

(2) $b^2 = a^2 + c^2 - 2ac \cos B$

(3) $c^2 = a^2 + b^2 - 2ab \cos C$

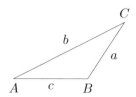

We have proven part 1 of the Law of Cosines. Part 3 is analogous, the only change is renaming sides and angles. Proving part 2 is slightly different, as $\angle B$ is larger than $90°$, but the approach of utilizing right triangles remains the same. We leave proving part 2 as an exercise.

As an interesting point of note, the Law of Cosines is a generalization of the Pythagorean Theorem. Consider part 2 with the oblique triangle given above. If $\angle B$ is reduced to $90°$, then triangle $\triangle ABC$ becomes right, which would result in part 2 of the Law of Cosines simplifying to the Pythagorean Theorem.

The Law of Cosines contains four variables, the lengths of all three sides, and one angle. You must know three of these variables to solve for the fourth. As a first example, consider the oblique triangle given in Figure 3. If we know the lengths of all

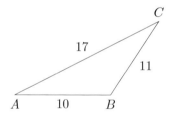

FIGURE 3. An oblique triangle in which all three sides are known.

three sides, we can find the measure of any angle.

$$11^2 = 17^2 + 10^2 - 2(17)(10) \cos A$$

$$17^2 = 11^2 + 10^2 - 2(11)(10) \cos B$$

$$10^2 = 11^2 + 17^2 - 2(11)(17) \cos C$$

Solving for any of the unknown angles involves some algebra, and the use of an inverse trigonometric function. Angle $\angle A$ would be

$$11^2 = 17^2 + 10^2 - 2(17)(10) \cos A$$

$$\frac{11^2 - 17^2 - 10^2}{-2(17)(10)} = \cos A$$

$$37.979° \approx A$$

Angles ∠B and ∠C could be found in exactly the same fashion.

Figure 4 illustrates our next scenario in which we know the lengths of two sides and the measure of the *included* angle. The included angle is the angle between the two known sides. The Law of Cosines allows us to solve for the length of the remaining

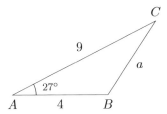

FIGURE 4. An oblique triangle in which two sides and the included angle are known.

side.

$$a^2 = 9^2 + 4^2 - 2(9)(4)\cos(27°)$$

$$a = \sqrt{96 - 72\cos(27°)}$$

$$a \approx 5.731$$

Finally, if we know the lengths of two sides and the measure of a *non-included* angle, we can still use the Law of Cosines to find the measurement of the third side. However, in this case there may be no solution, one solution, or two solutions. Consider our generic oblique triangle.

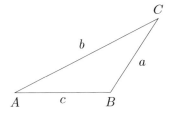

Suppose we know ∠A = 35°, a = 6, and b = 8. The Law of Cosines tells us that

$$6^2 = 8^2 + c^2 - 2(8)(c)\cos(35°)$$

$$0 = c^2 - 16\cos(35°)c + 28$$

which is a quadratic equation in standard form. The quadratic formula will allow us to solve for c.

$$c = \frac{-(-16\cos(35°)) \pm \sqrt{(-16\cos(35°))^2 - 4(1)(28)}}{2(1)}$$

After simplification, we have

$$c \approx 2.687 \quad \text{or} \quad c \approx 10.419$$

Thus, the initial stipulations presented in this example can be satisfied by two different triangles. Both are shown below.

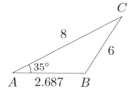
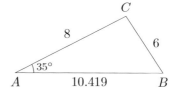

In similar problems involving two sides and a non-included angle, the number of solutions for the third side is as follows. In all cases, suppose we know a, b, and $\angle A$.

In the case where

$$a > b$$

then there is one solution for c, and triangle $\triangle ABC$ is oblique.

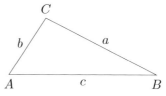

In the case where

$$\sin A = \frac{a}{b}$$

then there is one solution for c, and triangle $\triangle ABC$ is right. Here, $a = b\sin A$. This is the minimum possible length for a, otherwise a will be too short to connect and form a triangle.

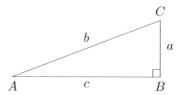

In the case where

$$a < b \sin A$$

there are no solutions as a is too short to connect and form a triangle.

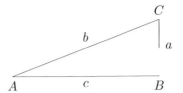

Finally, if

$$b \sin A < a < b$$

there are two solutions for c and two possible triangles that satisfy the given stipulations of a, b, and $\angle A$.

 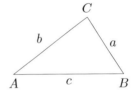

Exercises for Section 8.3.

(1) Prove part 2 of the Law of Cosines.

(2) The following exercises all make use of oblique triangles with angles A, B, C, and side lengths a, b, c. Solve for all unknown sides and angles.

(a) $A = 22°$, $b = 4$, $c = 1$

(b) $A = 15°$, $B = 138°$, $C = 27°$

(c) $A = 17°$, $a = 21$, $b = 9$

(d) $B = 31°$, $a = 8$, $b = 4$

(e) $A = 24°$, $a = 15$, $b = 33$

(f) $a = 4$, $b = 18$, $c = 20$

(3) You have decided to begin life anew as a land survey technician. Your first job finds you attempting to calculate the length of a lake. You hike to a point south of the lake, so that you can clearly see the Eastern and Western ends. This will be point P. At point P, you place a survey stake. You then

hike to the East and West ends of the lake, E and W respectively, placing a survey stake at each. The measured distance from P to E is 560 meters, and the measured distance from P to W is 872 meters. The measurement of the angle between the three survey stakes, $\angle EPW$, is $136°$. How long is the lake?

(4) The extreme world of land survey proved exhausting, so you have become a farmer. Your strawberry field is triangular, with side lengths 250, 135, and 350 meters. What is the area of your strawberry field?

8.4. The Law of Sines

With the Law of Cosines behind us, we can return our attention to discovering new relationships in oblique triangles. Consider triangle $\triangle ABC$ shown below.

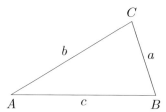

Suppose we draw a line from angle $\angle B$ to $\overline{AC}$ as follows

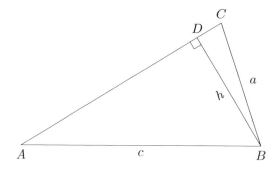

In right triangle $\triangle ABD$ we have

$$\sin A = \frac{h}{c} \quad \text{so} \quad h = c \sin A$$

While in right triangle $\triangle BCD$ we have

$$\sin C = \frac{h}{a} \quad \text{so} \quad h = a \sin C$$

Note that we have two different ways of expressing the variable h. Thus it must be true that

$$c \sin A = a \sin C$$

Some algebraic manipulation gives us

$$\frac{\sin A}{a} = \frac{\sin C}{c}$$

which is part of the relationship known as the *Law of Sines*.

The final piece of the law of sines can be found by starting with the same initial oblique triangle $\triangle ABC$ and drawing a line from angle $\angle C$ to $\overline{AB}$ as depicted below.

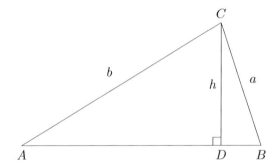

In right triangle $\triangle BCD$ we have

$$\sin B = \frac{h}{a} \quad \text{so} \quad h = a \sin B$$

While in right triangle $\triangle ACD$ we have

$$\sin A = \frac{h}{b} \quad \text{so} \quad h = b \sin A$$

Once again, we have two different expressions for the variable h. This means they must be equal.

$$a \sin B = b \sin A$$

Some algebra gives us

$$\frac{\sin B}{b} = \frac{\sin A}{a}$$

Combining this result with the previous leaves us with the full expression of the Law of Sines.

$$\frac{\sin A}{a} = \frac{\sin B}{b} = \frac{\sin C}{c}$$

The Law of Sines holds for all triangles, right or oblique. To illustrate an application, consider the oblique triangle shown in Figure 5. The lengths of the unknown sides can be quickly calculated as follows.

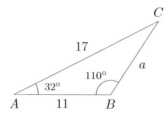

FIGURE 5. An example utilizing the Law of Sines

$$\frac{\sin 110°}{17} = \frac{\sin 32°}{a}$$

$$a\left(\sin 110°\right) = 17\left(\sin 32°\right)$$

$$a = \frac{17\left(\sin 32°\right)}{\sin 110°}$$

$$a \approx 9.586$$

An important word of caution; when solving an oblique triangle with two known sides and a non-included angle, the Law of Sines will not reveal the possibility that two solutions may exist (whereas the Law of Cosines will illustrate multiple solutions). Let us revisit a previous example that was solved via the Law of Cosines.

Recall that we know $\angle A = 35°$, $a = 6$, and $b = 8$.

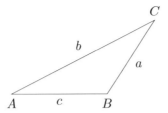

The Law of Sines allows us to write out the relationship

$$\frac{\sin 35°}{6} = \frac{\sin B}{8}$$

$$8(\sin 35°) = 6(\sin B)$$

$$\frac{8(\sin 35°)}{6} = \sin B$$

$$49.886 \approx B$$

As we now know angles $\angle A$ and $\angle B$, calculating $\angle C$ is straightforward.

$$\angle C \approx 180 - 35 - 49.886 \approx 95.114$$

Another application of the Law of Sines tells us

$$\frac{\sin 35°}{6} \approx \frac{\sin 95.114°}{c}$$

$$c(\sin 35°) \approx 6(\sin 95.114°)$$

$$c \approx \frac{6(\sin 95.114°)}{\sin 35°}$$

$$c \approx 10.419$$

This value for c is one of the two solutions we previously calculated with the Law of Cosines. Recall that the second solution was $c \approx 2.687$. Utilization of the Law of Sines in the non-included angle case does not reveal the existence of this second solution for c, and thus the other triangle that satisfies the initial stipulations of the problem.

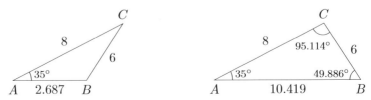

Our final example will utilize a *navigational heading*, referred to as a heading for short. A heading is simply the degrees, clockwise, from North. Traveling due North would be a heading of 0°, due East 90°, due South 180°, and due West 270°.

Suppose you are following a map to buried treasure. The map tells you the treasure is 9 kilometers due East from your current position, but there are many traps along the way. Rather than walk straight from your current position to the treasure, you plot a detour where you will hike at a 65° heading for some distance, then turn and hike at a 125° heading until you reach the buried treasure. What is the total distance of your detour to the treasure?

As always, begin with a clear picture of the scenario. The setup of the problem is given below in Figure 6, where the distances for each leg of your journey are referred to as L_1 and L_2.

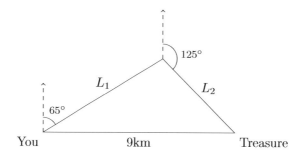

FIGURE 6. A setup of the buried treasure scenario.

The additional angles shown in Figure 7 come from subtraction (from either 90 or 180 degrees), and alternate interior angles.

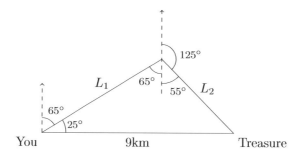

FIGURE 7. Determining the interior angles of the treasure scenario.

With these interior angles, we can easily calculate the remaining angle in the oblique triangle. The remaining angle would be

$$180° - 25° - 120° = 35°$$

Distances L_1 and L_2 would be given by

$$\frac{\sin 120}{9} = \frac{\sin 35}{L_1} \qquad\qquad \frac{\sin 120}{9} = \frac{\sin 25}{L_2}$$

$$L_1 (\sin 120) = 9 (\sin 35) \qquad\qquad L_2 (\sin 120) = 9 (\sin 25)$$

$$L_1 = \frac{9 (\sin 35)}{\sin 120} \qquad\qquad L_2 = \frac{9 (\sin 25)}{\sin 120}$$

$$L_1 \approx 5.96 \text{ km} \qquad\qquad L_2 \approx 4.391 \text{ km}$$

Thus, the total distance of the detour will be approximately

$$5.96 + 4.391 = 10.351 \text{ km}$$

Exercises for Section 8.4.

(1) Use the Law of Sines to solve problem 2 from the exercises in section 8.3. Compare your answers to the solutions you found using the Law of Cosines. What differences do you notice?

(2) Solve all triangles for all unknown angles and lengths.

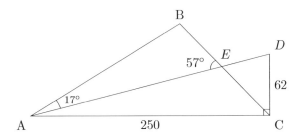

(3) Alex and Madison are walking East along a straight road. Alex is 6 km to the East of Madison. A mountain looms at the end of the road. From Alex's current position, he must look upwards at a 17° angle to see the top of the mountain. From Madison's current position, she must look upwards at a 11° angle to see the top of the mountain. How tall is the mountain? Disregard the height of Alex and Madison.

(4) You start racing dune buggies across the desert. During a long-distance race, you notice a sinkhole on the course in front of you. Rather than risk hitting the sinkhole, you veer off course (which was straight for the foreseeable future) at a heading of 320°. You continue along this heading for 3.2 km, and then take a heading of 32° back to the race course. How much extra distance did the detour add to your race?

(5) Spend some time and familiarize yourself with the definitions of Latitude and Longitude. Now, let us once again assume the Earth is a sphere with radius 6,378 km.

   (a) If you are on the equator (latitude of zero) and travel one degree of longitude, what distance have you covered?

   (b) If you are on a latitude of 27° N, and travel one degree of longitude, what distance have you covered?

   (c) If you are on a latitude of 82° S, and travel one degree of longitude, what distance have you covered? Ignore the fact that you are likely frozen.

CHAPTER 9

Answers To Exercises

This chapter contains answers to all exercises. As mentioned in the introduction, these answers are not your goal. The justification for these answers is your goal. Writing down an answer without understanding the "why" behind it is detrimental and a waste of your time. Using these answers to develop and solidify understanding is not.

1.2.1

(1) (a) $\frac{11}{30}$ (b) $\frac{300x^3}{y^3}$ (c) -2 (d) $\frac{1}{zn^3p^3}$

(2) $x = 3$ (3) $q = \frac{kp}{3p-k}$ (4) $x = \pm 1$

(5) $x = \frac{9}{2}$ or $x = -1$ (6) $x = 2 \pm \sqrt{3}$

(7) vertex: $(-2, 1)$, x-ints: $(-1, 0)$ and $(-3, 0)$, y-int: $(0, -3)$

(8) $\left(-\frac{1}{2}, \frac{4}{3}\right)$ (9) The distance is $\sqrt{29}$ units

2.1

(1) There are many possible answers for all parts of this question. One example for each part has been given below.
  (a) $(5, 1), (6, 4), (7, 3), (8, 2)$
  (b) $(1, 5), (2, 6), (3, 7), (4, 8)$
  (c) $(5, 1), (5, 2)$ This pairing scheme violates the definition in two ways. The element 5 is not uniquely paired, and domain elements $6, 7, 8$ have not been paired.

(2) (a) The domain of f is $\{1, 2, 5, 7, c\}$ and the range is $\{4, 3, 0, a, 9\}$.

(b) (i) 0
    (ii) 4
    (iii) Undefined
    (iv) 9
    (v) Undefined
    (vi) Undefined

(3) (a) $(3, 2), (6, 5), (9, 1), (12, 0)$
  (b) The domain of h is $\{3, 6, 9, 12\}$ and the range is $\{0, 1, 2, 5\}$.
  (c) $0 = h(12)$, $1 = h(9)$, $2 = h(3)$, $5 = h(6)$

(4) Answers will vary.

(5) The range is all elements of the codomain that have been paired.

(6) (a) The statement allows a domain element to pair with more than one unique range element.
  (b) The statement allows domain elements to remain unpaired.

2.2

(1) (a) k is onto, but not 1-1. Domain elements c and d both pair with range element h.
  (b) $g = k(a)$, $h = k(c)$, $i = k(b)$, $f = k(e)$, $h = k(d)$

(2) (a) g is 1-1, but not onto. Codomain element n has not been paired.
  (b) $m = g(2)$, $l = g(3)$, $p = g(1)$, $o = g(4)$

(3) (a) No. The codomain of f is the set T, and the codomain of g is the set U, and $T \neq U$.
  (b) The range of f is $\{a, b, c\}$.
  (c) The range of g is $\{a, b, c\}$.
  (d) Yes. f and g make all of the same pairs.

(4) Answers will vary.

(5) (a) There are several possible mappings.
     One example would be $(2, a), (4, b), (6, c), (8, d)$.
  (b) There are several possible mappings.
     One example would be $(2, a), (4, a), (6, b), (8, c)$
  (c) This is not possible given that A and B have the same number of elements. Fixing this would require removing an element from A, or adding an element to B. Suppose we remove 2 from A. One possible answer would be $(4, a), (6, b), (8, c)$.
  (d) This is not possible given that A and B have the same number of elements. Fixing this would require adding an element to A, or removing

an element from B. Suppose we add the element 10 to A. One possible answer would be $(2, a), (4, b), (6, c), (8, d), (10, d)$.

(6) (a) $(1, 2), (2, 4), (3, 6), (4, 8)$
   (b) Yes, f is 1-1. Given two distinct natural numbers, $n_1 \neq n_2$ we must have $f(n_1) \neq f(n_2)$
   (c) Yes, f is onto. Any even natural number, t, can be written as $f(n) = 2n = t$ where $n \in \mathbb{N}$.

(7) This must mean that S and T have the same number of elements. With regard to the previous problem, the function f is bijective. The sets $\mathbb{N}$ and T have the same number of elements.

2.3

(1) As f and g are both bijective, and the range of g is the domain of f, the composition $f \circ g$ will be a function. The composition $g \circ f$ cannot be a function as the range of f, set D, is not the domain of g.

(2) (a) Yes. Explanations will vary.
   (b) Yes. Explanations will vary.
   (c) No. Suppose we call $A = \{1, 3, a, t, 2\}$ and $B = \{b, 5, 2, 1, 9\}$. Then $g \circ f : B \to B$ and $f \circ g : A \to A$.

   (d) (i) t (iv) undefined (vii) undefined
       (ii) 3 (v) b (viii) 2
       (iii) 2 (vi) undefined   (ix) undefined

(3) Answers will vary. One possible example would be:

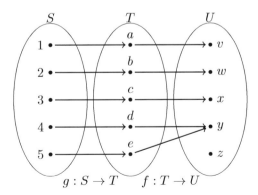

(4) The pairings are visible in the previous answer.

(5) No. If g is not onto, then f will be unable to pair all of its domain, the set T. This will violate the definition of a function.

2.4

(1) $(c, a), (j, 1), (z, 5), (0, 3), (v, l)$

(2) Sets will vary. In any case, f^{-1} is not a function as the elements in T that were not paired by f are now domain elements of f^{-1} that cannot be paired.

(3) Sets will vary. In any case, f^{-1} is not a function as there are elements in T, the domain of f^{-1} that are not uniquely paired.

(4) Yes. Explanations will vary.

(5)

x	$f(x)$	$f^{-1}(x)$
1	9	5
3	5	6
4	4	4
5	1	3
6	3	9
9	6	1

Full verification is left to the reader. To illustrate the process for the composition $f^{-1}(f(x))$ on the domain element 1:

$$f^{-1}(f(1)) = f^{-1}(9) = 1$$

Repeat this process for all elements in the domain of f. Then verify for the composition $f(f^{-1}(x))$.

3.1

(1) Independent variables are free to take on any value. Dependent variables will have their values dictated by the values of the independent variables. There are countless phenomena that are dependent on other, one example would be the fact that the temperature of Earth is dependent on the Earth's distance from the Sun (among other things).

(2) Any sketch of a non-vertical linear function clearly illustrates that the domain must be $\mathbb{R}$. For the range, as the function has a defined, non-zero slope, any vertical value will eventually be achieved.

(3) (a) -2 (c) -14 (e) $3a - 5$

    (b) -5 (d) $-\frac{7}{2}$ (f) $3\pi - 5$

(4) (a) 7 (c) $-\frac{7}{2}$ (e) $\frac{7}{2}(a-1)$

    (b) $\frac{7}{2}$ (d) $\frac{28}{5}$ (f) $\frac{7}{2}(\pi-1)$

(5) Sketches of graphs will vary. In any case, as the domain of a function is plotted on the horizontal axis, any two points on the same vertical line would correspond to a domain value with more than one unique pairing.

(6) Answers will vary. One example could be the pairings

$$(a, b), (c, d)$$

3.2

(1) (a) $\frac{36}{5}$ (b) 25 (c) $\frac{2\pi + 30}{5}$ (d) $x^2 + 6x + 9$

(2) This version of an equation pairs a single domain value, x, with infinite range values.

(3) Slope is the rate at which a linear function increases or decreases. A linear function will positive slope increases. A linear function with negative slope decreases. Slope is either positive or negative, so the function is either increasing or decreasing.

(4) (a) 81 (b) π^2 (c) 4 (d) 28561

(5) The constant, parabola, and absolute value functions are not 1-1. Any sketch will reveal that a horizontal line is capable of intersecting the graph of these functions in more than one location.

(6) (a) 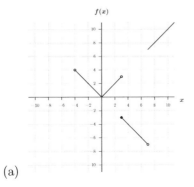 (b)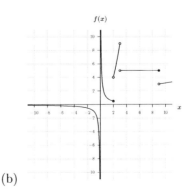

(7) $h(x) = -\frac{2}{3}x + 3$

(8) $f(x) = 2x + 3$

(9) (a) 2 (b) 4 (c) 6

3.3

(1) In all domains, $x \in \mathbb{R}$.

(a) $\{x \leq \frac{7}{9}\}$
(b) $\mathbb{R}$
(c) $\mathbb{R}$
(d) $\mathbb{R}$
(e) $\{x \leq 8 \text{ and } x \neq -8\}$
(f) $\{x \neq \frac{5}{2}\}$
(g) $\mathbb{R}$
(h) $\{x \geq 5\}$
(i) $\{x < -3 \text{ or } x > 2\}$
(j) $\{x \geq -5\}$
(k) $\{-2 \leq x \leq 2 \text{ or } x > 9\}$
(l) $\{x \neq -10\}$

(2) 2, 0, and -50 are in the range of f. No, f is not 1-1.

(3) Domain: $\{x \mid x \in \mathbb{R} \text{ and } -10 \leq x \leq -6 \text{ or } -4 \leq x \leq 1 \text{ or } -2 < x \leq 3\}$
Range: $\{f(x) \mid f(x) \in \mathbb{R} \text{ and } -2 \leq f(x) \leq \frac{1}{2} \text{ or } 2 \leq x \leq 9\}$
Yes, f is 1-1.

3.4

(1) The cubic and reciprocal functions are odd, and the absolute value function is even.

(2) No. A function having even and odd symmetry would have the relationship $f(-x) = -f(x) = f(x)$. This equality results in a contradiction when considering the sign of $f(x)$.

(3) $(7, 3)$

(4) (a) g is also even. (b) g is even.

(5) (a) h is also odd. (b) No.

(6) f is even, g is neither, h is odd.

(7) (a) 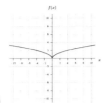 (b) 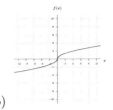 (c)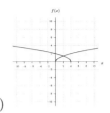

(8) (a) $f \neq g$, they have different domains.
(b) $f \neq g$, they have different domains.
(c) $f \neq g$, they pair $x < -1$ differently.
(d) $f = g$ as $x + a > 0$ and $x + b > 0$.

3.5

(1) (a) $j(x) = f(x) - 7$
(b) $k(x) = 9g(x)$
(c) $l(x) = -h(x)$
(d) $m(x) = 1 + \frac{2}{3}f(x)$
(e) $n(x) = \frac{1}{5}h(x) - 1$
(f) $p(x) = -\frac{7}{2}g(x) - 8$

(2) (a) $f(x) = \frac{1}{3}\left(\sqrt{x} + 9\right)$, $g(x) = \frac{1}{3}\left(\frac{1}{x} + 9\right)$, $h(x) = \frac{1}{3}\left(x^3 + 9\right)$
Compression after the shift, graphs will all appear to have been shifted up by 3.
(b) $f(x) = \frac{1}{3}\sqrt{x} + 9$, $g(x) = \frac{1}{3} \cdot \frac{1}{x} + 9$, $h(x) = \frac{1}{3}x^3 + 9$
Compression before shift, graphs will appear to have been shifted up by 9.

(3) (a) $f(x) = \sqrt{x}$. Compression by $\frac{5}{7}$, down 7.
(b) $f(x) = \frac{1}{x}$. Reflect over horizontal axis, up 6.
(c) $f(x) = x^3$. Stretch by $\frac{11}{2}$, up 3.
(d) $f(x) = |x|$. Stretch by 5, up $\frac{4}{9}$.
(e) $f(x) = x^2$. Up 1, stretch by $\frac{7}{3}$.
(f) $f(x) = \sqrt{x}$. Down 1, reflect over horizontal axis, stretch by $\frac{9}{8}$. Can also go down 1, stretch by $\frac{9}{8}$, then reflect.
(g) $f(x) = x^2$. Reflect over horizontal axis, compress by $\frac{6}{17}$, up 1. Can also compress by $\frac{6}{17}$, reflect, up 1.
(h) $f(x) = \frac{1}{x}$. Up 4, reflect over horizontal axis, stretch by $\frac{14}{3}$. Can also go up 4, stretch by $\frac{14}{3}$, then reflect.

3.6

(1) (a) $j(x) = f(x+12)$
    (b) $k(x) = g(x-9)$
    (c) $l(x) = h(-x)$
    (d) $m(x) = f\left(\frac{2}{5}x + 3\right)$
    (e) $n(x) = h\left(\frac{6}{13}x - 11\right)$
    (f) $p(x) = g(2x - 7)$

(2) (a) $f(x) = \sqrt{3x-9}$, $g(x) = \frac{1}{3x-9}$, $h(x) = (3x-9)^2$
    Compression after the shift, graphs will all appear to have been shifted right by 3.
    (b) $f(x) = \sqrt{3(x-9)}$, $g(x) = \frac{1}{3(x-9)}$, $h(x) = (3(x-9))^2$
    Compression before shift, graphs will appear to have been shifted right by 9.

(3) (a) $f(x) = \sqrt{x}$. Left 1, compression by $\frac{1}{4}$.
    (b) $f(x) = \frac{1}{x}$. Left 6, reflect over vertical axis.
    (c) $f(x) = x^3$. Right 9, stretch by 2.
    (d) $f(x) = |x|$. Right 2, compress by $\frac{1}{5}$.
    (e) $f(x) = x^2$. Left 4, compress by $\frac{2}{3}$.
    (f) $f(x) = \sqrt{x}$. Stretch by $\frac{7}{4}$, right 3.
    (g) $f(x) = x^2$. Reflect over vertical axis, right $\frac{7}{9}$.
    (h) $f(x) = \frac{1}{x}$. Reflect over vertical axis, left $\frac{6}{17}$.

3.7

(1) (a) $f(x) = \sqrt{x}$. **H:** Right 6. **V:** Compress by $\frac{2}{3}$.
    (b) $f(x) = \frac{1}{x}$. **H:** Left 7. **V:** Stretch by 5.
    (c) $f(x) = x^2$. **H:** Right 2, reflect across vertical axis. **V:** Compress by $\frac{1}{7}$, down 8.
    (d) $f(x) = \sqrt{x}$. **H:** Right 4, compress by $\frac{2}{3}$. **V:** Stretch by $\frac{3}{2}$, up 1.
    (e) $f(x) = x^2$. **H:** Right 1, compress by $\frac{1}{5}$. **V:** Compress by $\frac{4}{13}$, down 4.
    (f) $f(x) = \frac{1}{x}$. **H:** Right 11. **V:** Compress by $\frac{1}{7}$, reflect across horizontal axis, up 9. May also reflect, then compress, then up.
    (g) $f(x) = x^3$. **H:** Left 1, reflect across vertical axis. **V:** Stretch by 6, reflect across horizontal axis, up 3. May also reflect, then stretch, then up.
    (h) $f(x) = |x|$. **H:** Stretch by 6, left 2. **V:** Stretch by 12, up $\frac{11}{5}$.

(2) $g(x) = f(-x + 6)$

(3) The function m contains the following integer value points:

$$(0,1), (1,1), (2,2), (3,2), (4,1), (5,1), (6,0)$$

Rather than graph all transformations, we will transform these points and allow the reader to draw corresponding graphs.

(a) $(0,2), (-1,2), (-2,4), (-3,4), (-4,2), (-5,2), (-6,0)$
(b) $(-1,-1), (0,-1), (1,0), (2,0), (3,-1), (4,-1), (5,-2)$
(c) $(0, -\frac{1}{2}), (1, -\frac{1}{2}), (2, -1), (3, -1), (4, -\frac{1}{2}), (5, -\frac{1}{2}), (6, 0)$
(d) $(0,2), (\frac{1}{2}, 2), (1,3), (\frac{3}{2}, 3), (2,2), (\frac{5}{2}, 2), (3,1)$
(e) $(-6,1), (-3,1), (0,2), (3,2), (6,1), (9,1), (12,0)$
(f) $(0,-2), (-1,-2), (-2,-3), (-3,-3), (-4,-2), (-5,-2), (-6,-1)$

(4) (a) $f(x) = \sqrt{x-3} + 2$ (c) $f(x) = -\frac{1}{4}(x-4)^2 - 1$
    (b) $f(x) = \frac{1}{4}(x+2)^3 + 1$ (d) $f(x) = \frac{1}{x-3} - 1$

3.8

(1) (a) $\frac{1}{x-1}$ (c) $\frac{5x}{2}$ (e) $|x+5|$
    (b) $\sqrt{\frac{x+9}{x+2}}$ (d) $x^2 - 4x + 2$ (f) $x+6$

(2) (a) 2 (b) 2 (c) b (d) 0 (e) b (f) a

(3) (a) $f(g(x)) = \frac{1}{\sqrt{x-4}}$, $\{x \mid x \in \mathbb{R} \text{ and } x > 4\}$.
       $g(f(x)) = \sqrt{\frac{1-4x}{x}}$, $\{x \mid x \in \mathbb{R} \text{ and } 0 < x \leq \frac{1}{4}\}$.
    (b) $f(g(x)) = 8x - 4$, $\{x \mid x \in \mathbb{R} \text{ and } x \geq \frac{1}{2}\}$.
       $g(f(x)) = \sqrt{8x^2 - 1}$, $\{x \mid x \in \mathbb{R} \text{ and } x \leq -\sqrt{\frac{1}{8}} \text{ or } x \geq \sqrt{\frac{1}{8}}\}$.
    (c) $f(g(x)) = \frac{2x^2}{5+3x^2}$, $\{x \mid x \in \mathbb{R} \text{ and } x \neq 0\}$.
       $g(f(x)) = \frac{5x^2+30x+45}{4}$, $\{x \mid x \in \mathbb{R} \text{ and } x \neq -3\}$.
    (d) $f(g(x)) = \frac{\sqrt{9x^2-1}}{|x|}$, $\{x \mid x \in \mathbb{R} \text{ and } x \leq -\frac{1}{3} \text{ or } x \geq \frac{1}{3}\}$.
       $g(f(x)) = \frac{1}{9-x}$, $\{x \mid x \in \mathbb{R} \text{ and } x < 9\}$.

(4) (a) $j(x) = g(g(x))$ (c) $l(x) = h(g(x))$ (e) $n(x) = g(h(x))$
    (b) $k(x) = h(f(x))$ (d) $m(x) = f(f(x))$ (f) $p(x) = h(h(x))$

(5) $(0,9), (1,10), (4,8), (5,12), (6,11)$

3.9

(1) Sketches can be verified with graphing software. In all parts, the composition will be $p(q(x))$.
   (a) $p(x) = \sqrt{x}$, $q(x) = x^2 - 4$. Domain of $q : \{x \mid x \in \mathbb{R}\}$. Domain of $f : \{x \mid x \in \mathbb{R}$ and $-2 \le x \le 2\}$.
   (b) $p(x) = \frac{1}{x}$, $q(x) = x^2$. Domain of $q : \{x \mid x \in \mathbb{R}\}$. Domain of $g : \{x \mid x \in \mathbb{R}$ and $x \ne 0\}$.
   (c) $p(x) = |x|$, $q(x) = (x-1)^3$. Domain of $q : \{x \mid x \in \mathbb{R}\}$. Domain of $h : \{x \mid x \in \mathbb{R}\}$.
   (d) $p(x) = (x-1)^3$, $q(x) = |x|$. Domain of $q : \{x \mid x \in \mathbb{R}\}$. Domain of $j : \{x \mid x \in \mathbb{R}\}$.
   (e) $p(x) = x^2$, $q(x) = \sqrt{x}$. Domain of $q : \{x \mid x \in \mathbb{R}$ and $x \ge 0\}$. Domain of $k : \{x \mid x \in \mathbb{R}$ and $x \ge 0\}$.
   (f) $p(x) = \frac{1}{x}$, $q(x) = \sqrt{x}$. Domain of $q : \{x \mid x \in \mathbb{R}$ and $x \ge 0\}$. Domain of $l : \{x \mid x \in \mathbb{R}$ and $x > 0\}$.

(2)
$$\text{For } x \ge 0: \quad j(x) = (|x| - 1)^3 = (x-1)^3$$
$$\text{Thus for } -x: \quad j(-x) = (|-x| - 1)^3 = (x-1)^3$$

(3) Sketches can be verified with graphing software. In all parts, the composition will be $p(q(x))$ where $q(x) = |x|$

   (a) $p(x) = \sqrt{x+3}$
   (b) $p(x) = \frac{1}{x-4}$
   (c) $p(x) = |x - 3|$
   (d) $p(x) = (x-2) - 5$

(4) (a) 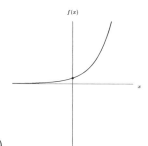 (b)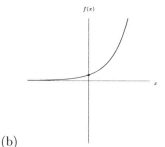

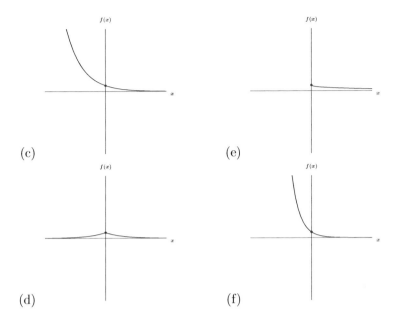

3.10

(1) For all domain and ranges, x, $f(x)$, and $f^{-1}(x) \in \mathbb{R}$.
  (a) Invertible domain has two options: $\{x \mid x \geq 5\}$, or $\{x \mid x \leq 5\}$
  Range: $\{f(x) \mid f(x) \geq -6\}$.
  The inverse depends on domain choice, respectively $f^{-1}(x) = \sqrt{x+6}+5$
  or $f^{-1}(x) = -\sqrt{x+6}+5$
  Domain of inverse: $\{x \mid x \geq -6\}$
  Range depends on choice for invertible domain, respectively $\{f^{-1}(x) \mid f^{-1}(x) \geq 5\}$, or $\{f^{-1}(x) \mid f^{-1}(x) \leq 5\}$.
  (b) Invertible domain has two options: $\{x \mid x \geq -2\}$, or $\{x \mid x \leq -2\}$
  Range: $\{f(x) \mid f(x) \geq -3\}$.
  The inverse depends on domain choice, respectively $f^{-1}(x) = \sqrt{x+3}-2$, or $f^{-1}(x) = -\sqrt{x+3}-2$
  Domain of inverse: $\{x \mid x \geq -3\}$
  Range depends on choice for invertible domain, respectively $\{f^{-1}(x) \mid f^{-1}(x) \geq -2\}$, or $\{f^{-1}(x) \mid f^{-1}(x) \leq -2\}$
  (c) D: $\{x \mid x \neq -3\}$, R: $\{f(x) \mid f(x) \neq -5\}$. $f^{-1}(x) = \frac{2}{x+5} - 3$, D: $\{x \mid x \neq -5\}$, R: $\{f^{-1}(x) \mid f^{-1}(x) \neq -3\}$.
  (d) D: $\mathbb{R}$, R: $\mathbb{R}$. $f^{-1}(x) = \sqrt[3]{x-7}+5$, D: $\mathbb{R}$, R: $\mathbb{R}$.
  (e) D: $\{x \mid x \leq 3\}$, R: $\{f(x) \mid f(x) \leq 2\}$. $f^{-1}(x) = 3 - (x-2)^2$, D: $\{x \mid x \leq 2\}$, R: $\{f^{-1}(x) \mid f^{-1}(x) \leq 3\}$.

(f) Invertible domain has two options: $\{x \mid x \geq 3\}$, or $\{x \mid x \leq -3\}$
Range: $\{f(x) \mid f(x) \geq 0\}$.
The inverse depends on domain choice, respectively $f^{-1}(x) = \sqrt{x^2 + 9}$, or $f^{-1}(x) = -\sqrt{x^2 + 9}$
Domain of inverse: $\{x \mid x \geq 0\}$
Range depends on choice for invertible domain, respectively $\{f^{-1}(x) \mid f^{-1}(x) \geq 3\}$, or $\{f^{-1}(x) \mid f^{-1}(x) \leq -3\}$

(2) Verify with graphing software.

(3) (a) Number of chewed trees per acre when there are two beavers per acre.
(b) Twelve beavers per acre results in thirty five chewed trees per acre.
(c) Number of beavers per acre when there are fifty four chewed trees per acre.
(d) Twenty nine chewed trees per acre results in ten beavers per acre.

(4) The result is x. (5) Explanations will vary.

(6) $a = \frac{1}{6}$, $b = -\frac{13}{6}$

(7)

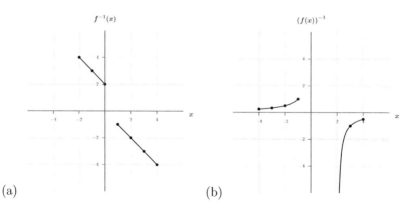

(a) (b)

(8)

x	$f(x)$	$(f(x))^{-1}$	$f(-x)$	$f^{-1}(x)$
-3	2	$\frac{1}{2}$	-1	-2
-2	-3	$-\frac{1}{3}$	0	1
-1	3	$\frac{1}{3}$	-2	3
0	1	1	1	2
1	-2	$-\frac{1}{2}$	3	0
2	0	UN	-3	-3
3	-1	-1	2	-1

4.1

(1) (a) $y = 0.916x + 1.878$

(b) Yes, $r^2 \approx 0.84$. Roughly 84% of the variation in $f(x)$ is explained by the regression line.

(c)

x	$f(x)$	Residual
2	5	1.289
5	6	-0.459
7	9	0.707
10	8	-3.041
12	12	-0.873
15	18	2.377

(d) The sum of the residuals is zero (rounding error is the result of any non-zero residual sum). Discussions will vary.

(e) $(11, 11.954)$, $(5.591, 7)$

(2) $y = -0.852x + 34.045$ where $r^2 = 0.938$. Given such a high coefficient of determination, the model appears quite good. The prediction values would be $(10, 25.52)$, $(12, 23.816)$, $(20, 17)$ where all output values are in °C. A standard room temperature of $25°C$, would correspond to a cooling time of 10.616 seconds. As our data extends beyond this value, we must be in a situation where the room temperature is much lower than standard. We need more information to be able to extrapolate. In this case, the scope of the model should not extend beyond our observed values.

(3) $y = 101.28x - 21.513$. The slope of this model suggests that roughly 101 IU of vitamin D are produced for every minute of sunlight exposure. The vertical axis intercept would mean that approximately 21 IU of vitamin D are consumed for zero minutes of sunlight. This would need to be verified by an additional study, the vertical axis intercept has no applicable meaning for this data. The model fits the data well, $r^2 = 0.97$. The high end scope of the model is somewhat hard to determine. It would require additional information about the point at which the body terminates vitamin D production. There is literature suggesting that 10,000 IU of production is realistic, so we can use that as the extreme scope value. This IU value would correspond to a time of 98.523 minutes of sun exposure. The scope of the model would is then $0 < x < 98.523$.

4.2

(1) (a) min $(5, 22.667)$
      max $(25, 156)$

(b) min $(11.744, 0.197)$
    no max

(2) (a) $h(0)$ is the height of the ball at $t = 0$. This means the cannon is 1 meter off the ground. This makes sense.
   (b) $\{t \mid t \in \mathbb{R} \text{ and } 0 \leq t \leq 7.66\}$
   (c) Height of 144.495 meters at $t = 3.872$ seconds.

(3) You should study 52.568 hours per week.

(4) (a) $(2.5, \sqrt{2.5})$.   (b) Length of 8, width of 2.

(5) The width of the rectangle is 0.98 meters, the length of the rectangle is 1.96 meters, and the radius of the circle is 0.98 meters.

(6) Aim for the road at a point 1.945 kilometers to the west of your car.

(7) (a) $C(x) = 12 - 0.5x$
   (b) $R(x) = x(12 - 0.5x) + 240$
   (c) 12 people

(8) Start on the fiber optic cable at the point directly south of the house belonging to Friend 1. Move along the cable a distance of 9.658 meters. This is where the three of you connect.

(9) The length of the large pen is 50 meters, the width of the large pen is 20 meters. Each of the small pens is 12.5 meters in length, 20 meters in width.

(10) The radius of the can is 3.291 centimeters, the height is 10.268 centimeters.

4.3

(1) (a) $9,219.41
    (b) $9,219.41
    (c) Interest is compounded at the end of each year, not during the year.

(2) (a) $16,858.21
    (b) $17,036.15
    (c) Interest is compounded daily, so the amount at four and a half years would be larger than the amount at four years.

(3) 22.11%   (4) $W = \frac{Pr(1+r)^n}{(1+r)^n - 1}$ (5) $(3.003 \times 10^{-7})A_\circ$

(6) (a) $A(N) = A_\circ (0.73)^N$ where N is the thickness of the glass in cm.
    (b) Answers vary.

(7) (a) $A(N) = 2^N$
    (b) 8 : 19 am
    (c) 2 minutes
    (d) Colonizing new planets won't save us from overpopulation.

(8) 317.97 mg

(9) Discussions will vary. In order, the four models are

$$A_\circ \left(1 + \frac{0.032}{12}\right)^{12N} \quad A_\circ \left(1 + \frac{0.032}{52}\right)^{52N} \quad A_\circ \left(1 + \frac{0.032}{365}\right)^{365N} \quad A_\circ e^{0.032N}$$

5.1

(1) Graphs can be verified with graphing software.

(2) (a) $f(x) = 2^x$, $(-1, \frac{1}{2})$ and $(1, 2)$. **H:** Left 7, and reflect over vertical axis. $(-8, -\frac{1}{2})$ and $(-6, -2)$.
    (b) $f(x) = e^x$, $(-1, \frac{1}{e})$ and $(1, e)$. **V:** Compress by $\frac{3}{5}$, **H:** Reflect over vertical axis. $(1, \frac{3}{5e})$ and $(-1, \frac{3e}{5})$.
    (c) $f(x) = \frac{1}{2}^x$, $(-1, 2)$ and $(1, \frac{1}{2})$. **H:** Left 4, stretch by 2. $(-10, 2)$ and $(-6, \frac{1}{2})$.

(d) $f(x) = 5^x$, $(-1, \frac{1}{5})$ and $(1, 5)$. **V:** Stretch by 4, **H:** Left 3, reflect over vertical axis. $(4, \frac{4}{5})$ and $(2, 20)$.

(e) $f(x) = \frac{9^x}{5}$, $(-1, \frac{5}{9})$ and $(1, \frac{9}{5})$. **V:** Reflect over horizontal axis, stretch by 6, down 2, **H:** Left 1, stretch by $\frac{11}{2}$. $(-11, -17)$ and $(0, -\frac{74}{5})$.

(f) $f(x) = e^x$, $(-1, \frac{1}{e})$ and $(1, e)$. **V:** Stretch by e, down e, **H:** Rotate over vertical axis. $(1, 1-e)$ and $(-1, e^2 - e)$.

(3) (a) $f(x) = \left(\frac{1}{2}\right)^x + 3$ (b) $g(x) = -3^x + 2$

(4) Domain of f and g is $\mathbb{R}$. Yes, $f = g$.

(5) g is even, f is odd.

5.2

(1) Explanations will vary. All should focus on the fact that the base must be greater than zero.

(2) (a) $\frac{1}{2} \log_5 x + 4 \log_5 y$ (d) $\frac{1}{2} \log x + \log y - 3 \log z$
(b) $4 \ln x + \frac{1}{3} \ln y - \frac{1}{2} \ln z$ (e) $\log_b x + \frac{3}{4} \log_b y - \frac{5}{4} \log_b z$
(c) $2 \log_7 2 - \log_7 x - \log_7 y$ (f) $3 \log_a 3 + 6 - 4 \log_a 2 - 2 \log_a y$

(3) (a) $\log_2 \left(\frac{x^{\frac{1}{2}}}{y^3 z^4} \right)$ (c) $\log_3 \left(\frac{x-4}{x+4} \right)$

(b) $\log \left(\frac{x^3}{y^{\frac{4}{3}} z^5} \right)$ (d) $\log_4 \left(\frac{x+2}{x+3} \right)$

(4) (a) $2^3 = 8$ (c) $e^1 = e$ (e) $7^0 = 1$
(b) $\frac{1}{3}^{-1} = 3$ (d) $a^M = r$ (f) $10^{-2} = 0.01$

(5) (a) $\log_3 27 = 3$ (c) $\log_{36} 6 = \frac{1}{2}$ (e) $\log_{\frac{2}{3}} \frac{3}{2} = -1$
(b) $\ln 1 = 0$ (d) $\log_{125} 5 = \frac{1}{3}$ (f) $\log_B P = r$

(6) (a) $x = \frac{1}{2}$ (b) $x = 25$ (c) $x = 4^{\frac{7}{3}}$ (d) $x = 8^3$

(7) (a) **D:** $\mathbb{R}$, **R:** $\{f(x) \mid f(x) \in \mathbb{R} \text{ and } f(x) > 0\}$. $f^{-1}(x) = \frac{\ln x}{3}$, **D:** $\{x \mid x \in \mathbb{R} \text{ and } x > 0\}$, **R:** $\mathbb{R}$.

(b) **D:** $\{x \mid x \in \mathbb{R} \text{ and } x > 0\}$, **R:** $\mathbb{R}$. $g^{-1}(x) = \frac{1}{e^x}$, **D:** $\mathbb{R}$, **R:** $\{g^{-1}(x) \mid g^{-1}(x) \in \mathbb{R} \text{ and } g^{-1}(x) > 0\}$.

(c) **D:** $\{x \mid x \in \mathbb{R} \text{ and } x > -5\}$, **R:** $\mathbb{R}$. $h^{-1}(x) = 2^x - 5$, **D:** $\mathbb{R}$, **R:** $\{h^{-1}(x) \mid h^{-1}(x) \in \mathbb{R} \text{ and } h^{-1}(x) > -5\}$.

(d) **D:** $\mathbb{R}$, **R:** $\{j(x) \mid j(x) \in \mathbb{R} \text{ and } j(x) > 0\}$. $j^{-1}(x) = \log_3 x + 4$, **D:** $\{x \mid x \in \mathbb{R} \text{ and } x > 0\}$, **R:** $\mathbb{R}$.

(8) All compositions are the identity map on the domain of the inner function.

(9) (a) $10^{-pH} = H^+$
   (b) Given the exponential relationship established in the previous part of this question, higher pH values result in a decrease of hydrogen ions, which means the solution is more basic.
   (c) The solution with a pH of 4 is 100 times more acidic than the solution with a pH of 6.

5.3

(1) All sketches can be verified with graphing software
   (a) Transformations on $f(x) = \ln x$. $\left(\frac{1}{e}, -1\right)$ and $(e, 1)$. **H:** left 1. **V:** reflection across x-axis. $\left(\frac{1-e}{e}, 1\right)$ and $(e-1, -1)$.
   (b) Composition, $g(x) = l(m(x))$ where $l(x) = \ln x$ and $m(x) = \frac{1}{x}$.
   (c) Transformations on $f(x) = \ln x$. $\left(\frac{1}{e}, -1\right)$ and $(e, 1)$. **H:** right 3. **V:** reflection across x-axis and compress by $\frac{1}{4}$, up 3. $\left(\frac{1+3e}{e}, \frac{13}{4}\right)$ and $\left(e+3, -\frac{11}{4}\right)$.
   (d) Transformations on $f(x) = \log x$. $\left(\frac{1}{10}, -1\right)$ and $(10, 1)$. **H:** reflection across vertical axis. **V:** stretch by 2, up 7. $\left(-\frac{1}{10}, 5\right)$ and $(-10, 9)$.
   (e) Composition $k(x) = f(g(x))$ where $f(x) = \ln\left(\frac{1}{x}\right)$ and $g(x) = |x|$. k is even.
   (f) Composition. $l(x) = f(g(x))$ where $f(x) = \log x$ and $g(x) = x^4 + 1$. l is even.

(2) f is odd.

5.4

(1) (a) $x = \frac{-3\ln 4}{\ln 4 - 1}$
   (b) $x = \sqrt[7]{14}$
   (c) $x = \pm\sqrt{\ln 13}$
   (d) $x = \frac{3\ln 7 - \ln 9}{2\ln 7 - \ln 9}$
   (e) $x = -2 + \sqrt{6}$
   (f) $x = 1$, $x = \frac{1}{e^2}$, $x = e^2$
   (g) $x = \pm\sqrt{\frac{\ln 3 + \ln 5}{\ln 5}}$
   (h) $x = \frac{\ln 5}{\ln 4}$

(2) (a) $k = \ln\left(\frac{Q+R}{Bt}\right)$
   (b) $w = Ln^P$
   (c) $Z = \frac{J}{B + e^2}$
   (d) $H = \sqrt[\frac{\log R}{S}]{D^4 + F}$
   (e) $r = \frac{e^{\frac{Pe}{a}} - B}{D}$
   (f) $a = \frac{1}{P}\ln\left(\frac{NU}{\ln b - mU}\right)$

(3) (a) **D:** $\{x \mid x \in \mathbb{R} \text{ and } x > -1\}$, **R:** $\mathbb{R}$. $f^{-1}(x) = \sqrt{e^x} - 1$. **D:** $\mathbb{R}$, **R:** $\{f^{-1}(x) \mid f^{-1}(x) \in \mathbb{R} \text{ and } f^{-1}(x) > -1\}$.
(b) **D:** $\mathbb{R}$, **R:** $\{g(x) \mid g(x) \in \mathbb{R} \text{ and } g(x) > 0\}$. $g^{-1}(x) = \frac{1}{5}\ln\left(\frac{x}{3}\right)$. **D:** $\{x \mid x \in \mathbb{R} \text{ and } x > 0\}$, **R:** $\mathbb{R}$.
(c) The function is not invertible on its defined domain. The selected modification is **D:** $\{x \mid x \in \mathbb{R} \text{ and } x \geq 0\}$, **R:** $\{h(x) \mid h(x) \in \mathbb{R} \text{ and } h(x) > 0\}$. $h^{-1}(x) = \sqrt{\log(4x)}$. **D:** $\{x \mid x \in \mathbb{R} \text{ and } x > 0\}$, **R:** $\{h^{-1}(x) \mid h^{-1}(x) \in \mathbb{R} \text{ and } h^{-1}(x) \geq 0\}$.
(d) The function is not invertible on its defined domain. The selected modification is **D:** $\{x \mid x \in \mathbb{R} \text{ and } x > \sqrt{7}\}$, **R:** $\mathbb{R}$. $j^{-1}(x) = \sqrt{e^x + 7}$. **D:** $\mathbb{R}$, **R:** $\{j^{-1}(x) \mid j^{-1}(x) \in \mathbb{R} \text{ and } j^{-1}(x) > \sqrt{7}\}$.

(4) The models are given by

$$A(N) = A_0 \left(\frac{1}{2}\right)^{\frac{N}{63}} \quad \text{and} \quad F(t) = A_0 e^{\left(\frac{\ln\frac{1}{2}}{63}\right)t}$$

Regardless of your choice, approximately 84.7% of the initial amount remains after 15 years.

6.1

(1) (a) $f(x) \to \infty$ as $x \to \infty$, $f(x) \to -\infty$ as $x \to -\infty$.
(b) $g(x) \to -\infty$ as $x \to \infty$, $g(x) \to -\infty$ as $x \to -\infty$.
(c) $h(x) \to -\infty$ as $x \to \infty$, $h(x) \to \infty$ as $x \to -\infty$.
(d) $j(x) \to \infty$ as $x \to \infty$, $j(x) \to \infty$ as $x \to -\infty$.

(2) (a) $x = 0$ crosses, $x = 3$ does not cross, $x = -2$ crosses, $x = -11$ does not cross.
(b) $x = -4$ crosses, $x = -2$ crosses, $x = -3$ crosses.
(c) $x = 7$ crosses, $x = \frac{-3 \pm \sqrt{5}}{2}$ crosses both.
(d) $x = -2$ does not cross, $x = 3$ crosses.

(3) (a) $f(x) = (x-2)(x-4)(x+3)$
(b) $f(x) = \frac{1}{4}(x)(x+4)$
(c) $f(x) = \frac{1}{2}(x+1)(x-1)(x-2)$
(d) $f(x) = \frac{1}{5(3-a)}(x)(x+2)(x-a)$

(4) $f(x) = x^3 - 9x^2 + 21x - 5$ (5) $f(x) = \frac{7}{3}x^3 - 6x^2 + \frac{2}{3}x + 7$

(6) (a) $f(x) = \frac{1}{2}(x+3)(x-2)^2$

(b) $f(x) = (x-1)(x+2)^2(x+1)$

6.2

(1) (a) $f(x) \to \frac{9}{4}$ as $x \to \pm\infty$.
   (b) $g(x) \to 0$ as $x \to \pm\infty$.
   (c) $h(x) \to \infty$ as $x \to \infty$, $h(x) \to -\infty$ as $x \to -\infty$.
   (d) $j(x) \to 2x + 3$ as $x \to \pm\infty$.

(2) (a) $f(x) \to 1$ as $x \to \pm\infty$, hole at $x = -1$, vertical asymptote at $x = 2$, intercepts are $(-3, 0)$ and $\left(0, -\frac{3}{2}\right)$.
   (b) $g(x) \to 2x - 3$ as $x \to \pm\infty$, hole at $x = 0$, vertical asymptote at $x = 2$, intercepts are $(-1, 0)$ and $(4, 0)$.
   (c) $h(x) \to 3$ as $x \to \pm\infty$, hole at $x = -2$, vertical asymptotes at $x = 2$ and $x = 3$, intercepts are $(0, 0)$ and $(6, 0)$.
   (d) $j(x) \to -\frac{1}{2}x + 1$ as $x \to \pm\infty$, no holes, vertical asymptote at $x = -2$, intercepts are $\left(0, \frac{1}{2}\right)$, $(-\sqrt{2}, 0)$ and $(\sqrt{2}, 0)$.

(3) (a) $\left(\frac{1}{12}, 1\right)$ (b) $\left(\frac{15}{14}, -\frac{27}{14}\right)$

6.3

(1) (a) Power, polynomial.   (d) Exponential.
   (b) Power, rational.   (e) Power.
   (c) Power, rational.   (f) Rational.

(2) f will have a larger output. The intersection where f ultimately overtakes g is approximately $(13.884, 1{,}070{,}446.587)$.

(3) (a) $F(d) = \frac{m_1 m_2 G}{d^2}$
   (b) 1.261×10^{-15} N
   (c) $15{,}627.321$ m
   (d) Assuming Earth has diameter 6.378×10^6 m, and mass 5.972×10^{24} kg, the force in Newtons will end up as $m_1(9.796)$ where m_1 is the mass of any object on the surface of the Earth.

(4) (a) $k = \frac{1}{3^{\frac{\ln 5}{\ln 4}}}$, $p = \frac{\ln 5}{\ln 4}$ (b) $k = \frac{3}{2^{\frac{\ln 3}{\ln \frac{5}{2}}}}$, $p = \frac{\ln 3}{\ln \frac{5}{2}}$

7.1

(1) (a) 45°

(d) 72°

(g) −288°

(b) −25.714°

(e) 720°

(h) −32.727°

(c) 143.239°

(f) −60°

(i) 315°

(2) (a) Sine is positive in quadrants one and two, cosine is positive in one and four.

(b) (i) $\sin(0) = 0$
     $\cos(0) = 1$

 (ii) $\sin(-\frac{\pi}{2}) = -1$
     $\cos(-\frac{\pi}{2}) = 0$

 (iii) $\sin(\pi) = 0$
      $\cos(\pi) = -1$

 (iv) $\sin(\frac{3\pi}{2}) = -1$
      $\cos(\frac{3\pi}{2}) = 0$

 (v) $\sin(2\pi) = 0$
     $\cos(2\pi) = 1$

 (vi) $\sin(\frac{\pi}{2}) = 1$
      $\cos(\frac{\pi}{2}) = 0$

 (vii) $\sin(-\pi) = 0$
       $\cos(-\pi) = -1$

 (viii) $\sin(-\frac{3\pi}{2}) = 1$
        $\cos(-\frac{3\pi}{2}) = 0$

(3) $\sin\theta \approx 0.936$

(4) $\cos\theta \approx \pm 0.897$

(5) $\theta \approx 3.793$

(6) (a) Approximately $3,498.542$ cm.
   (b) Approximately 0.028
   (c) The distance to the ramp will also double, the sine of the angle will remain the same. Discussions will vary.

7.2

(1) The full unit circle is provided.

(2) (a) $\frac{\sqrt{2}}{2}$ (d) 0 (g) $\frac{\sqrt{3}}{2}$ (j) 1
   (b) $-\frac{1}{2}$ (e) $\frac{1}{2}$ (h) -1 (k) $\frac{\sqrt{2}}{2}$
   (c) 0 (f) $-\frac{1}{2}$ (i) 0 (l) $-\frac{\sqrt{3}}{2}$

(3) Questions will vary.

7.3

(1) Tangent, secant and cosecant are

   (a) $0, -1$, undefined
   (b) $-\sqrt{3}, -2, \frac{2\sqrt{3}}{3}$
   (c) undefined, undefined, 1
   (d) $0, 1$, undefined
   (e) $1, -\sqrt{2}, -\sqrt{2}$
   (f) $\sqrt{3}, -2, -\frac{2\sqrt{3}}{2}$
   (g) undefined, undefined, -1
   (h) $-1, \sqrt{2}, -\sqrt{2}$

(2) Sketches will vary.

(3) Secant is even, cosecant is odd, cotangent is odd.

(4) There cannot be a point on the unit circle where the vertical coordinate of the point is larger than one.

(5) (a) 0.784 (c) -1.612 (e) 0.62 (g) 0.784
   (b) -1.264 (d) 1.275 (f) 0.62 (h) -0.784

7.4

(1) Discussions will vary.

(2) Graphs can be verified with graphing software. For sine, the zeros are $(-\pi, 0), (0, 0), (\pi, 0)$, turning points are $\left(-\frac{\pi}{2}, -1\right), \left(\frac{\pi}{2}, 1\right)$, and the inflection point is $(0, 0)$. For cosine, the zeros and inflection points are $\left(-\frac{\pi}{2}, 0\right), \left(\frac{\pi}{2}, 0\right)$, and the turning point is $(0, 1)$.

(3) (a) $\frac{-\sqrt{6}-\sqrt{2}}{4}$ (c) $\frac{(\sqrt{2}-\sqrt{6})^2}{4}$ (e) $\sqrt{6}-\sqrt{2}$

(b) $\frac{\sqrt{3}}{2}$ (d) $-\frac{2\sqrt{3}}{3}$ (f) $\frac{(\sqrt{2}-\sqrt{6})(\sqrt{6}-\sqrt{2})}{4}$

(4) Discussions will vary.

7.5

(1) Graphs can be verified with graphing software.

(2) (a), (b), (c), (d)

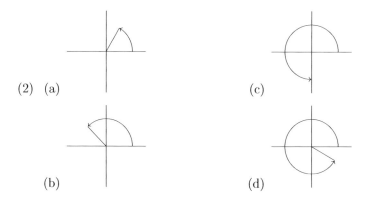

(3) No. The function f can pair domain values of $x = k\pi$ where $k \in \mathbb{Z}$, while the function g cannot.

7.6

(1) Graphs can be verified with graphing software.

(2) All transformations are vertical. Selected points will vary, and can be verified with substitution into the transformed function. Graphs can be verified with graphing software.

  (a) Stretch by a factor of 2, down 1. **D:** $\mathbb{R}$. **R:** $\{f(x) \mid -3 \le f(x) \le 1\}$.
  (b) Compression by a factor of $\frac{1}{3}$, up 3. **D:** $\mathbb{R}$. **R:** $\{g(x) \mid \frac{7}{3} \le g(x) \le \frac{10}{3}\}$.
  (c) Reflection across horizontal axis, down 2. **D:** $\{x \mid x \in \mathbb{R} \text{ and } x \ne \frac{\pi}{2} + k\pi \text{ where } k \in \mathbb{Z}\}$. **R:** $\mathbb{R}$.
  (d) Stretch by a factor of 2, down 2. **D:** $\{x \mid x \in \mathbb{R} \text{ and } x \ne k\pi \text{ where } k \in \mathbb{Z}\}$. **R:** $\{j(x) \mid j(x) \in \mathbb{R} \text{ and } j(x) \le -4 \text{ or } j(x) \ge 0\}$.
  (e) Compression by a factor of $\frac{1}{2}$, reflection across horizontal axis, or vice versa. **D:** $\{x \mid x \in \mathbb{R} \text{ and } x \ne \frac{\pi}{2} + k\pi \text{ where } k \in \mathbb{Z}\}$. **R:** $\{k(x) \mid k(x) \in \mathbb{R} \text{ and } k(x) \le -\frac{1}{2} \text{ or } k(x) \ge \frac{1}{2}\}$.
  (f) Stretch by a factor of π, up 1. **D:** $\{x \mid x \in \mathbb{R} \text{ and } x \ne k\pi \text{ where } k \in \mathbb{Z}\}$. **R:** $\mathbb{R}$.

(g) Stretch by a factor of $\frac{\pi}{2}$. **D:** $\mathbb{R}$. **R:** $\{m(x) \mid -\frac{\pi}{2} \leq m(x) \leq \frac{\pi}{2}\}$.

(h) Compression by a factor of $\frac{2}{\pi}$. **D:** $\mathbb{R}$. **R:** $\{n(x) \mid -\frac{2}{\pi} \leq n(x) \leq \frac{2}{\pi}\}$.

(3) $f(x) = 3\pi \sin x + e$

7.7

(1) Graphs can be verified with graphing software. Domains and ranges are presented in interval notation.
  (a) **Horiz:** Right $\frac{\pi}{3}$. **Vert:** Comp $\frac{2}{7}$. **Period:** 2π. **Phase:** $\frac{\pi}{3}$. **Amp:** $\frac{2}{7}$. **Equil:** $y = 0$. **D:** $\mathbb{R}$. **R:** $\left[-\frac{2}{7}, \frac{2}{7}\right]$.
  (b) **Horiz:** Reflection. **Vert:** Stretch π, down π. **Period:** 2π. **Phase:** None. **Amp:** π. **Equil:** $-\pi$. **D:** $\mathbb{R}$. **R:** $[-2\pi, 0]$.
  (c) **Horiz:** Left 2π. **Vert:** Down 1. **Period:** π. **Phase:** NA **Amp:** NA **Equil:** NA **D:** $x \neq \frac{\pi}{2} + k\pi$ **R:** $\mathbb{R}$
  (d) **Horiz:** Left $\frac{\pi}{4}$. **Vert:** Reflection. **Period:** π **Phase:** $-\frac{\pi}{8}$. **Amp:** 1. **Equil:** $y = 0$. **D:** $\mathbb{R}$. **R:** $[-1, 1]$.
  (e) **Horiz:** Right 3π, stretch by 4. **Vert:** Reflection, up 2π. **Period:** 8π. **Phase:** 12π. **Amp:** 1. **Equil:** $y = 2\pi$. **D:** $\mathbb{R}$. **R:** $[2\pi - 1, 2\pi + 1]$.
  (f) **Horiz:** Left $\frac{\pi}{2}$, reflection, stretch by $\frac{3}{2}$. **Vert:** None. **Period:** $\frac{3\pi}{2}$. **Phase:** NA. **Amp:** NA. **Equil:** NA. **D:** $x \neq \frac{3\pi}{2}k$. **R:** $\mathbb{R}$.
  (g) **Horiz:** Compression by $\frac{1}{e}$. **Vert:** Reflection. **Period:** $\frac{2\pi}{e}$. **Phase:** None. **Amp:** 1. **Equil:** $y = 0$. **D:** $\mathbb{R}$. **R:** $[-1, 1]$.
  (h) **Horiz:** Right 1, stretch by π. **Vert:** None. **Period:** $2\pi^2$. **Phase:** π. **Amp:** 1. **Equil:** $y = 0$. **D:** $\mathbb{R}$. **R:** $[-1, 1]$.

(2) $f(x) = 12 \cos \left(\frac{2\pi}{3}(x - \pi)\right) - 2$

7.8

(1) The restrictions on sine and cosine fail as they do not result in a 1-1 function, and they also restrict the range. The restriction on tangent does not restrict the range of tangent, but the points $(0, 0)$ and $(\pi, 0)$ prevent this restriction from being 1-1.

(2) (a) $\left(\frac{\pi}{4}\right)$ and $\left(\frac{3\pi}{4}\right)$
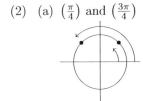

(b) $\left(\frac{\pi}{3}\right)$ and $\left(\frac{5\pi}{3}\right)$
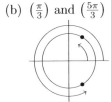

(c) $\left(\frac{\pi}{6}\right)$ and $\left(\frac{11\pi}{6}\right)$

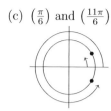

(e) $\left(\frac{\pi}{6}\right)$ and $\left(\frac{5\pi}{6}\right)$

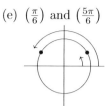

(d) $\left(\frac{4\pi}{3}\right)$ and $\left(\frac{5\pi}{3}\right)$

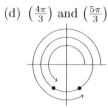

(f) $\left(\frac{3\pi}{4}\right)$ and $\left(\frac{5\pi}{4}\right)$

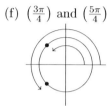

(3) Interpreting the tangent function as pairing an arc with the slope of the line between the origin and the arc endpoint, the inverse tangent would have a domain of all possible slope values, and a range of corresponding arcs on the unit circle.. The domain of the inverse would be $\mathbb{R}$, all possible values of slope. The range would be the non-inclusive right or left half of the unit circle (as the arcs on the right or left half will produce all possible values for the slope of the line connecting the arc endpoint and the origin). The choice of $\left(-\frac{\pi}{2}, \frac{\pi}{2}\right)$ would correspond to the right half of the unit circle.

(4) In all solutions, $k \in \mathbb{Z}$.
   (a) $x \approx \pm 0.722 + 2\pi k$.
   (b) $x = \pi k$.
   (c) $x = \frac{\pi}{3} + \pi k$

(5) This relationship is not true. There are a variety of possible ways to explain. Illustrating that the domains of these three inverse functions do not agree immediately invalidates the relationship, choosing a particular value and showing that the equation is false immediately invalidates the relationship, and graphing inverse tangent and the ratio of inverse sine and cosine will also clearly illustrate that the equation is false.

(6) There are many possible domain restrictions for each function. Domain and ranges are given in interval notation.
   (a) Domain restriction $\left[0, \frac{\pi}{2}\right]$. Range $[-1, 1]$. $f^{-1}(x) = \frac{\cos^{-1}(x)}{2}$.
   (b) Domain restriction $\left[-\frac{7\pi}{2}, -\frac{5\pi}{2}\right]$. Range $[-2, 2]$. $g^{-1}(x) = \sin^{-1}\left(\frac{x}{2}\right) - 3\pi$.
   (c) Domain restriction $\left(-\frac{\pi}{8}, \frac{\pi}{8}\right)$. Range $\mathbb{R}$. $h^{-1}(x) = \frac{\tan^{-1}(x-1)}{4}$.
   (d) Domain restriction $\left[-\frac{\pi}{2}, \frac{\pi}{2}\right]$. Range $[-2, 0]$. $j^{-1}(x) = \sin^{-1}(x+1)$.

7.9

(1) Answers will vary. In all cases, the sawtooth graph is a result of the domain restrictions that were requited to invert sine and cosine.

(2) (a) $\frac{2\sqrt{2}}{3}$ (b) $\frac{\sqrt{21}}{5}$ (c) $\frac{\sqrt{2}}{2}$ (d) Undefined

(3) For both f and g, the domain is $[-1, 1]$ and the range is $[0, 1]$.

(4) An arc of length 0.929 has a horizontal endpoint coordinate that is equal to the vertical endpoint coordinate for an arc of length 2.5.

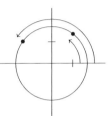

(5) (a) $\sqrt{-4x^2 - 4x}$ (b) $\frac{\sqrt{-x^2-8x-15}}{x+4}$

(6) (a) $x = 3$ (b) $x = \frac{1}{2}$

7.10

(1) (a) $\left(\frac{3}{2}, \frac{3\sqrt{3}}{2}\right)$ (b) $\left(6\cos\left(-\frac{5}{6}\right), 6\sin\left(-\frac{5}{6}\right)\right)$

(2) $\left(r\cos\left(\frac{A}{r}\right), r\sin\left(\frac{A}{r}\right)\right)$ (3) Approximately 92.759 km.

(4) The first part of this answer comes from the realization that the plane will travel $4,500$ km on a circle of radius $6,389$ km, which corresponds to an approximate distance of $4,492.252$ km on the surface of the Earth. From there, this problem can be approached the same way as the previous. The tunnel path is approximately 100.031 km shorter.

8.1 For all answers, $k \in \mathbb{Z}$.

(1) (a) $x \approx -0.545 + \frac{4\pi}{3}k$ or $x \approx 1.212 + \frac{4\pi}{3}k$. On the interval, $x \approx 1.212$, $x \approx 3.643$, $x \approx 5.401$.

(b) $x = -\frac{5\pi}{12} + 5\pi k$ or $x = \frac{35\pi}{12} + 5\pi k$. There are no solutions on the interval.

(c) $x \approx 0.474 + k$. Solutions on the interval are for $k = 0$ through $k = 5$.

(d) $x = \frac{\pi}{6} + \frac{\pi}{3}k$. Solutions on the interval are for $k = 0$ through $k = 2$.

(2) (a) $-\frac{\pi}{12} + 2\pi k \leq x \leq \frac{7\pi}{12} + 2\pi k$.

(b) $\frac{\pi}{4} + 2\pi k < x < \frac{5\pi}{4} + 2\pi k$.

(c) $\frac{5\pi}{6} + 2\pi k < x < \frac{13\pi}{6} + 2\pi k$.

(d) $-1 - \frac{\pi}{6} + 2\pi k < x < -1 + \frac{\pi}{6} + 2\pi k$, or $1 - \frac{5\pi}{6} + 2\pi k < x < 1 + \frac{5\pi}{6} + 2\pi k$.

(3) (a) $x = \pi k$, or $x = \frac{\pi}{6} + 2\pi k$, or $x = \frac{5\pi}{6} + 2\pi k$.

(b) $x \approx \pm 0.806 + \frac{2\pi}{3}k$, or $x = \frac{2\pi}{3}k$.

(c) $x \approx \pm 0.804 + 2\pi k$.

(d) $x = \pi k$.

(e) $x \approx 0.107 + \pi k$, or $x \approx -0.658 + \pi k$.

(f) No solution.

8.2 Proofs for problems 1-3 are left to the reader. For all answers, $k \in \mathbb{Z}$.

(4) (a) $x = \pm \frac{2\pi}{3} + 2\pi k$, or $x = 2\pi k$.

(b) $x = \pi k$.

(c) $x = \frac{\pi}{4} + \pi k$.

(d) $x = \pi k$, or $x = \frac{3\pi}{4} + \pi k$.

(e) $x = \frac{\pi}{2} + 2\pi k$.

8.3

(1) Proof is left to the reader.

(2) Numeric values are approximate. Yours may be slightly different due to rounding differences.

(a) $a = 3.095$, $b = 4$, $c = 1$, $A = 22°$, $B = 151.049°$, $C = 6.951°$.

(b) Cannot be solved.

(c) $a = 21$, $b = 9$, $c = 29.441$, $A = 17°$, $B = 7.178°$, $C = 155.802°$.

(d) No triangle satisfies these conditions.

(e) Two possible solutions:
$a = 15$, $b = 33$, $c = 36.843$, $A = 24°$, $B = 63.485°$, $C = 92.515°$.
$a = 15$, $b = 33$, $c = 23.45$, $A = 24°$, $B = 116.515°$, $C = 39.485°$.

(f) $a = 4$, $b = 18$, $c = 20$, $A = 10.475°$, $B = 54.9°$, $C = 114.624°$.

(3) The lake is approximately $1,331$ meters long.

(4) The area of the field is approximately $13,255$ square meters.

8.4

(1) Answers are the same with the exception that the second solution for the case with two triangles is not identified.

(2) $\angle ABC = 106°$, $\angle DAC = 13.928°$, $\angle ACE = 43.072°$, $\angle AEC = 123°$, $\angle ADC = 76.072°$, $\angle DCE = 46.928°$, $\angle DEC = 57°$. $\overline{AD} = 257.573$, $\overline{AB} = 177.61$, $\overline{BC} = 133.668$, $\overline{AE} = 203.571$, $\overline{CE} = 71.753$, $\overline{DE} = 54.003$.

(3) The mountain is approximately 3.202 km tall.

(4) The detour added approximately 1.339 km to your race.

(5) (a) 111.317 km.   (b) 99.184 km.   (c) 15.492 km.

Made in the USA
Middletown, DE
10 April 2020

88784729R00152